EIGHTH EDITION

LABORATORY CHEMISTRY FOR THE HEALTH SCIENCES

GEORGE I. SACKHEIM
DENNIS D. LEHMAN

PRENTICE HALL Upper Saddle River, NJ 07458

Senior Editor: John Challice
Production Editor: Dawn Blayer
Supplement Cover Designer: PM Workshop Inc.
Special Projects Manager: Barbara A. Murray
Supplement Cover Manager: Paul Gourhan
Associate Editor: Mary Hornby
Manufacturing Buyer: Ben Smith

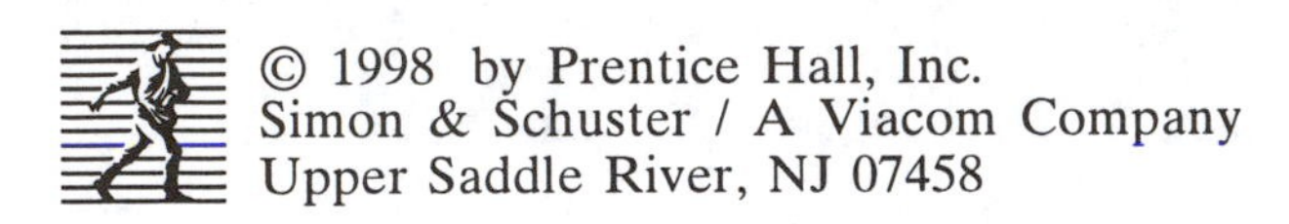

Simon & Schuster / A Viacom Company
Upper Saddle River, NJ 07458

Printed in the United States of America

10 9 8 7 6 5 4 3 2 1

ISBN 0-13-757717-6

Prentice Hall International (UK) Limited, *London*
Prentice Hall of Australia Pty. Limited, *Sydney*
Prentice Hall Canada, Inc., *London*
Prentice Hall Hispanoamericana, S.A., *Mexico*
Prentice Hall of India Private Limited, *New Delhi*
Prentice Hall of Japan, Inc., *Tokyo*
Simon & Schuster Asia Pte. Ltd., *Singapore*
Editora Prentice-Hall do Brazil, Ltda., *Rio de Janeiro*

Contents

Note to the Student

This laboratory manual is designed to

1. provide you with some concrete examples of various concepts presented in your textbook of chemistry
2. provide some practical applications of chemistry
3. allow you to gain confidence in your ability to work as an individual
4. provide an opportunity to work on multiconceptual problems

The experiments are planned to be done individually. Each experiment contains a prelab study quiz to ensure that you understand information before actually doing the lab work. The experiments will contain questions related to the work done. Sometimes these questions can be answered from the experiment itself; at other times you will need to go to an outside reference.

To introduce you to the practical use of laboratory work, several clinical case histories have been included in this manual. These case histories and their corresponding laboratory results are the basis for questions on (1) clinical applications and (2) scientific aspects. Although you may not be able to answer all the questions, you should receive valuable insight into the direct applications of laboratory work.

Note to the Instructor

The chemistry laboratory should provide your students with a unique opportunity: the chance to learn firsthand about the theories and laws of science that are so important to everyday life. Historically, it has been from laboratories—many of them not so elaborate as the one in which you work with your students—that great scientific progress has been made to benefit mankind.

The laboratory experiments in this manual were chosen to reinforce basic concepts and to provide settings in which these concepts might be brought together. We have tried to allow room for changes by individual instructors in many of the experiments.

Since carbon tetrachloride, benzidine, chloroform, and benzene are now considered to be carcinogens, their use was discontinued.

Although this book was written for use with the authors' textbook, *Chemistry for the Health Sciences* (8th ed., 1997), it can also be employed with many similar texts. A teacher's manual with lists of chemicals, equipment, and suggestions is also available from the publisher.

Most of the experiments are designed for individual work for a two- or three-hour lab period. Students should be able to the prelab study questions after reading the introduction to each experiment. Some questions at the end of the experiment may require outside reference material.

To introduce the students to the practical use of laboratory work, several clinical case histories have been included in this manual. These case histories and their corresponding laboratory results are the basis for questions on (1) clinical applications and (2) scientific aspects. Although beginning students may not be able to answer all the questions, they should receive a valuable insight into the direct applications of laboratory work.

Safety and Procedures in the Chemistry Laboratory

Safety and Procedures in the Chemistry Laboratory

SAFETY

All people who work in chemistry laboratories should be aware of certain inherent dangers that exist. Although the utmost precautions are taken by the authors of laboratory textbooks, instructors, and laboratory assistants, it is "the nature of the beast" that should keep you aware of the possibility of accidents. The chemistry laboratory should provide you with hours that are both academically stimulating and enjoyable.

Care should be taken by each student to follow certain basic rules of safety that will greatly minimize, if not eliminate, the dangers inherent to any chemistry course. It should be your responsibility to commit these safety rules to memory and practice them at all times. They are as follows:

1. Wear protective glasses *at all times* in the laboratory. In most states it is against the law not to wear them while in the lab.
2. Carefully read each experiment before coming to the laboratory.
3. Pay particular attention to the cautions mentioned in this book.
4. Be attentive to all verbal instructions given by your teacher and laboratory assistant.
5. Be considerate of the well-being of others in the laboratory as well as of yourself.
6. Enjoy yourself as you perform the experiments, but do not create, or tolerate from others, an atmosphere of "fooling around."
7. Note the following specific causes of many minor laboratory accidents and do everything in your power to avoid them:
   (a) Tendency of students to touch glassware that is hot.
   (b) Failure to follow instructions when inserting glass in a rubber stopper or cork.
   (c) Failure to be aware of the presence of others at all times.
   (d) Failure to clean up broken glassware immediately.
   (e) Tendency to taste or smell chemicals. *Do not taste anything* in the laboratory. To smell a gas or vapor, use your cupped hand to bring a small sample of the gas or vapor to your nose.

By avoiding these dangers, you will be doing yourself a favor and will derive maximum benefit from this course. In case of an accident should occur, no matter how slight, it should be reported immediately to the instructor.

Keep in mind that literally every accident that takes place in a course such as this one can be avoided by being prepared and following directions concerning your personal safety.

Find the location of the following safety features in the laboratory:

1. Fire extinguisher
2. Laboratory shower
3. Fire blanket
4. First aid kit
5. Eyewash
6. Disposal units (both chemical and paper)
7. Paper towels

Apparatus

1. Keep your equipment and work area clean.
2. At the end of the laboratory period, clean up your area with a wet paper towel.
3. Replace broken equipment as soon as possible.
4. Do not borrow equipment from others. If you need extra equipment, obtain it from the stockroom.

Chemicals

1. Laboratory reagents are shared by several students. Leave them in their assigned area, unless instructed to take them to your bench.
2. Some liquids may be disposed of by pouring them down the sink; others may not. (These include most organic solvents.) Ask before you pour anything into the sink.
3. Solids should be discarded in designated waste containers. Do not pour them into the sink.
4. Bottle stoppers should be always be held and never put down on the bench top.
5. Try to avoid removing excessive amounts of a reagent, but *never* return excess reagent to the reagent container.

PROCEDURES

The Gas Burner

1. The most commonly used gas burners are the Bunsen or Tirrill type. They are similar in that they both have an air adjustment on the burner; they differ in that the Tirrill type also has a gas adjustment screw at the bottom. (See Figure 1.) Observe the gas and air adjustments on the burner as well as the gas adjustment on the laboratory bench.

2. Light the burner as follows. Close the air port. Attach the burner hose to the gas supply and open the valve all the way. Bring a lighted match to the top of the burner from the side rather than from above. Adjust the air supply until a double bluish cone appears. The gas supply valve may have to be closed slightly.
3. The temperature range of the flame is indicated in Figure 2. Observe the differences in temperature within the flame by inserting a piece of iron wire into various parts of the flame and comparing the amount of time it takes for the wire to glow.

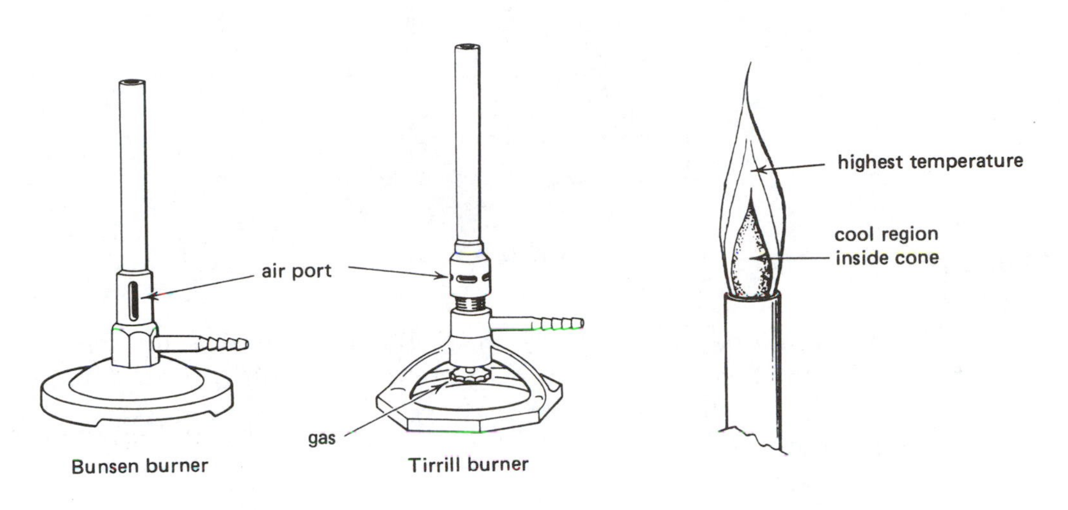

FIGURE 1

FIGURE 2

Heating Methods

1. To heat a test tube over a flame, grasp the tube with a test tube holder and move the tube back and forth across the flame (Figure 3). Be careful not to point the tube at anyone, yourself included, while it is being heated. Perform this procedure with a test tube filled with water.

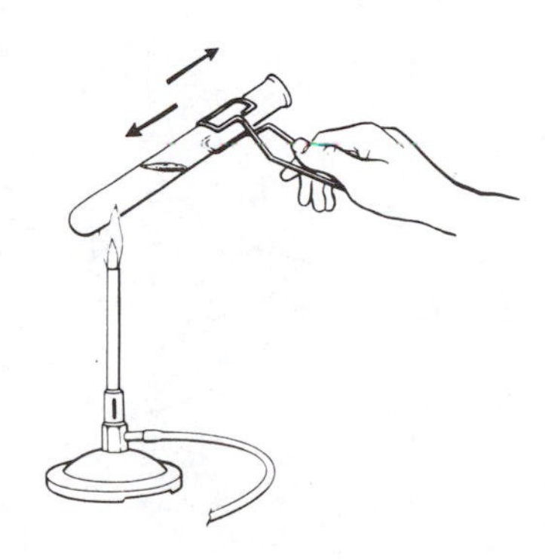

FIGURE 3

2. To heat a test tube that is mounted to a ring stand, hold the burner and pass the flame over the area to be heated (Figure 4). It is not necessary to perform this operation at this time.

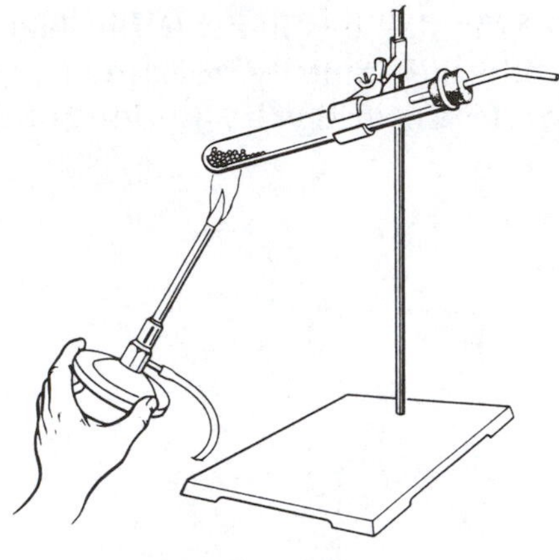

FIGURE 4

3. When heating beakers or flasks, place them on an asbestos screen in order to distribute the heat evenly (Figure 5). Be sure that the iron ring that supports the container is tightly screwed to the ring stand. It is not necessary to perform this operation at this time.

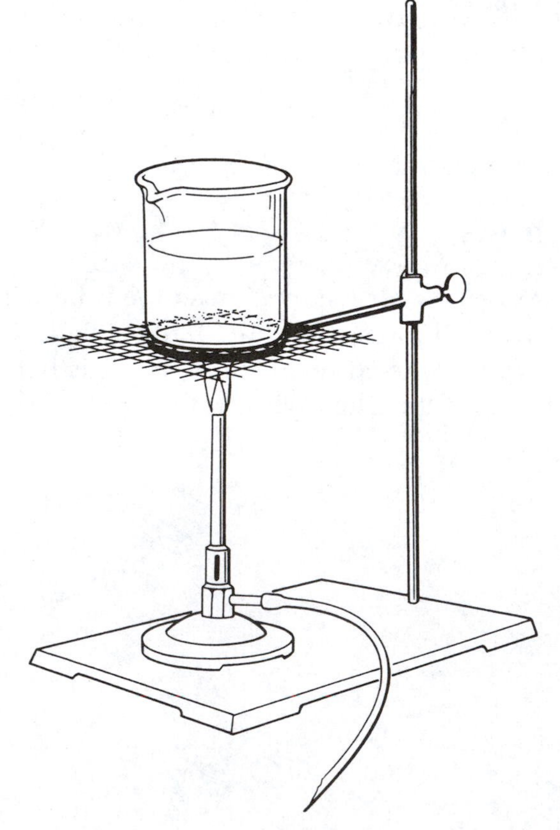

FIGURE 5

Glass-Working Methods

1. To break glass tubing correctly, scratch the glass with one firm stroke, using the edge of a file. Place the thumbs together on the side opposite the scratch and quickly break the glass as you would snap a twig (Figure 6).

   Prepare three pieces of 6-mm glass tubing 15 cm in length and two pieces of 4-mm solid glass rod also 15 cm in length. Do not discard these sections.

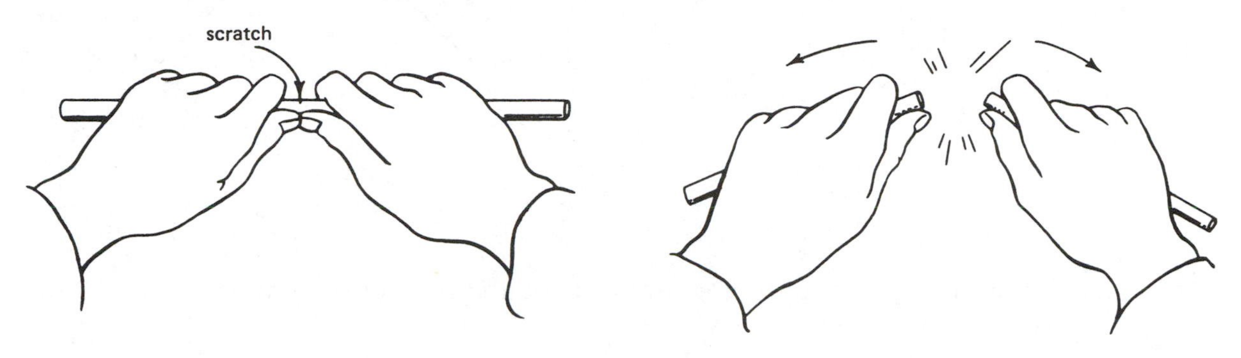

FIGURE 6

2. To fire polish glass means to smooth off the sharp ends on glass that has been cut. This may be done by rotating the rough end in the hottest part of a flame (Figure 7).

   Fire polish the five pieces of glass that you have just prepared. Allow sufficient time for the hot glass to cool before handling.

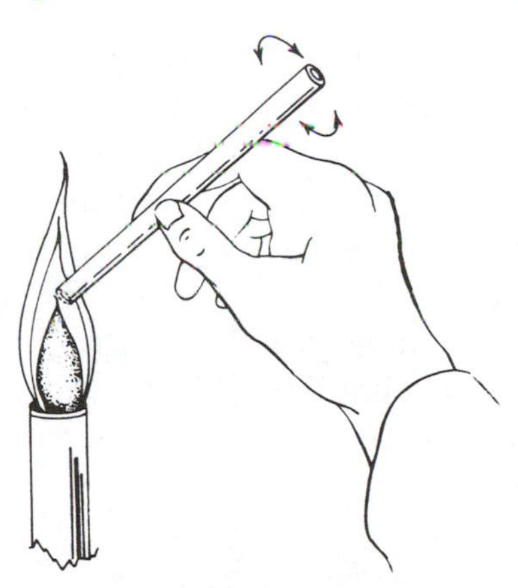

FIGURE 7

3. To bend glass tubing properly, place a wing-top on your burner. Hold the glass horizontally in the flame (Figure 8, page 8).

   Rotate the tubing back and forth until the glass becomes soft and bends under its own weight.

   Remove the tubing from the flame and bend to the desired angle. (Figure 9, page 8).

   Do not set the tubing down until it hardens. Allow ample time for cooling before handling it again.

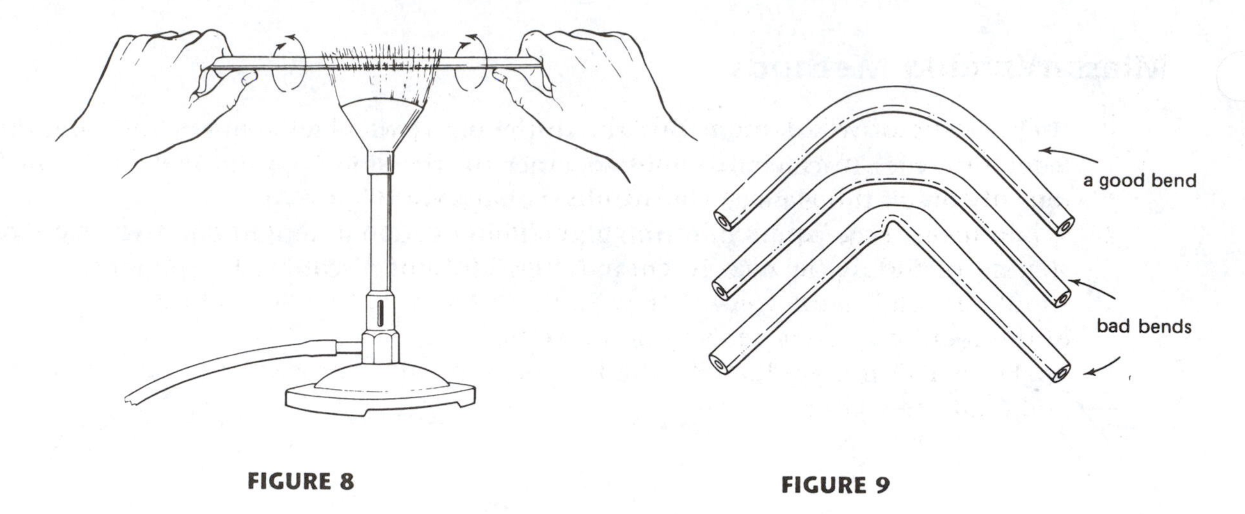

FIGURE 8

FIGURE 9

Bend the three pieces of tubing that you cut and fire polished in parts 1 and 2 of this section as follows:

(a) 45° bend at the midpoint of the tubing.
(b) 90° bend at the midpoint of the tubing.
(c) 90° bend 5 cm from the end of the tubing.

(Save all pieces of glass prepared in this experiment. You will have use for them later in the course.)

4. To insert a piece of glass tubing safely through a rubber stopper, fire polish the tubing, cool, and moisten that part of the tubing that is to pass through the stopper. Lubricate the inside of the hole in the stopper with glycerol (or water) and hold the stopper (Figure 10) so that, if the tubing should break, it will not be directed toward your hand. With your hands held close together, rotate the tubing back and forth and push it gently through the stopper (Figure 10). It is suggested that you hold the tubing with a towel to guarantee a good grip. Insert a piece of polished 6-mm tubing 10 cm long through a rubber stopper until approximately 2 cm comes through the stopper.

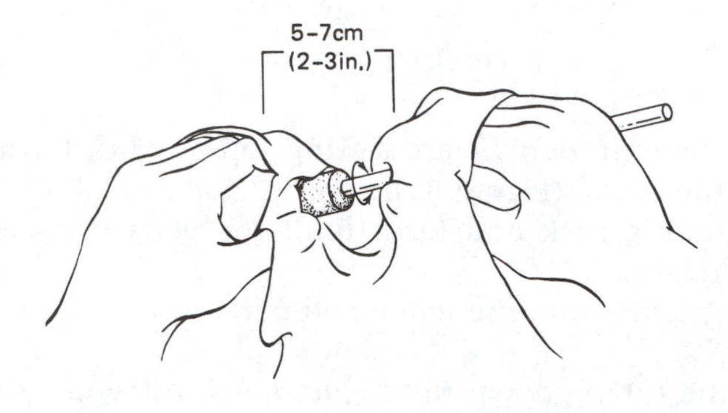

FIGURE 10

Measurements

1. Your instructor will demonstrate the use of the types of balances in your laboratory. If time permits at the end of the experiment, make several practice weighings of common objects such as coins, stoppers, and so on.
2. When measuring liquid volumes in graduated cylinders or burets, note that the surface of the liquid will be curved. The bottom of this curve (the meniscus) should be read as the volume (Figure 11). Make sure that your eye is at the level of the meniscus when reading liquid volumes.

   Fill a 100-mL graduated cylinder approximately three fourths full of water and read the volume.

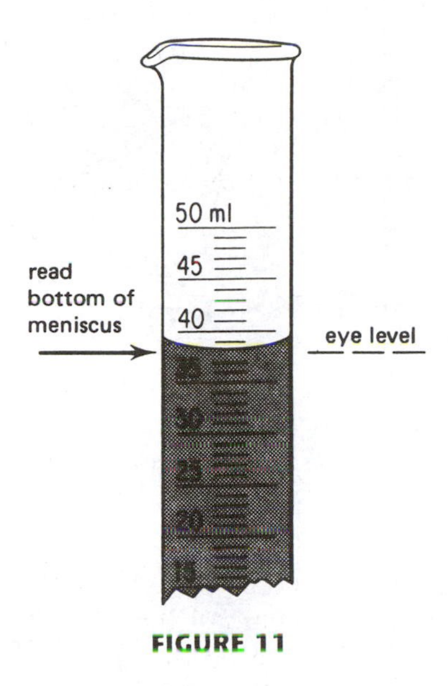

FIGURE 11

Filtration

Filtration is a process whereby solid particles are removed from a liquid by pouring the mixture through filter paper. To prepare a filter, fold the filter paper in half. Again fold the paper in half, and then separate the folds so that you have one thickness of paper on one side and three thicknesses of paper on the other side (Figure 12).

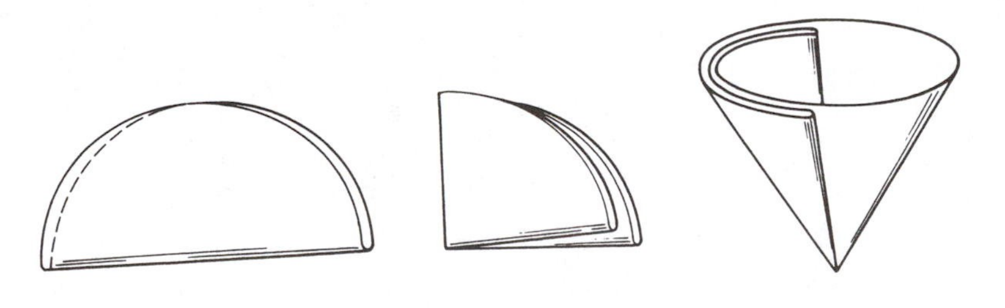

FIGURE 12

Prepare a piece of filter paper as described previously. Place the paper in a funnel and moisten the paper with water (very slightly) so that it will remain in the funnel. Using the apparatus as pictured in Figure 13, filter 25 mL of a mixture of charcoal and water. (Prepare the misture by adding 1 g of charcoal to 25 mL of water and stirring.)

FIGURE 13

Inorganic Chemistry Exercises

Name ______________________ **Class** __________ **Room** __________

Exercise 1
The Metric System

INTRODUCTION

Every day we encounter a wide variety of materials. When we come across a new substance, we attempt to describe and classify it according to its properties. Measurements are involved in determining many properties and play an important part in chemistry and medicine. The metric system of measurement is used in all scientific work. Although the United States is only slowly converting to this system, we shall use it exclusively in this manual.

There are two advantages to the metric system. First, it resembles our monetary system in that multiples and fractions based on the number 10 are used. The important prefixes are listed in Table 1-1. Second, there are relatively few basic units needed. The unit of length is the meter; the unit of volume is the liter; the unit of mass (weight) is the gram.

TABLE 1-1

Prefix	Abbreviation	Decimal Value
micro-	μ	0.000001
milli-	m	0.001
centi-	c	0.01
deci-	d	0.1
kilo-	k	1000

Sometimes we combine measurements to express a relationship between them. Examples include feet per second (ft/sec), miles per gallon (mi/gal), and cents per kilowatt-hour (c/kWhr). One important property of a substance, density, is defined in terms of two types of units: weight (mass) and volume. That is, density may be defined as mass (weight) per unit volume and is generally expressed mathematically as:

$$\text{density} = \frac{\text{mass}}{\text{volume}}$$

Example: A piece of gold has a volume of 4.5 mL and has a mass of 87 g. What is its density?

$$\text{density} = \frac{\text{mass}}{\text{volume}} = \frac{87\text{g}}{4.5\text{ mL}} = 19\,\frac{\text{g}}{\text{mL}}$$

PRELAB STUDY QUIZ

1. How many millimeters are in a meter? ______________________

2. The metric unit of volume is the ______________________.

3. There are ______________________ pounds in a kilogram.

4. If a 20-mL block of aluminum has a mass of 54 g, what is the density of aluminum?

SPECIAL HAZARD

Naphthalene is an irritant and is flammable. Hydrocarbons are highly flammable and their vapors may be explosive in the air. Be very careful with flames.

PROCEDURE

A. Length

Record all data on the Data Record.

1. Measure and record your height (in inches) using a yard stick.
2. Measure and record your height (in centimeters) using a meter stick.
3. Measure and record the length of a test tube in inches using an English ruler.
4. Measure and record the length of the same test tube using a meter stick.
5. Divide the centimeter reading by the inch reading in both of the preceding measurements and compare with the value reported in a conversion table.

B. Weight (mass)

1. Weigh an unknown specimen to the nearest 0.1 g (decigram). Save for Part C.
2. Weigh a piece of filter paper to the nearest 0.01 g (centigram).
3. Weigh a clean, dry 250-mL beaker to the nearest 0.1 g. Save this beaker for Part D.

C. Volume

Use a graduated cylinder for making the following volume measurements.

1. Fill a 250-mL beaker to the 250-mL mark with water; measure the volume of water to the nearest milliliter, using a graduated cylinder.
2. Put about 20 mL of water in the graduated cylinder. Record the volume. Tilt the graduated cylinder and slowly slide the preweighed unknown specimen into it. Be careful not to spill any water. Return the graduated cylinder to an upright position and record the new volume. The difference in volumes is the volume of the specimen. (Note that the specimen must be completely submerged.)

D. Density

1. Calculate the density of the unknown specimen.
2. Using a graduated cylinder add exactly 100 mL of water (or alcohol) to the 250-mL beaker from Part B. Reweigh the beaker and the liquid. Calculate the density of the liquid.

DATA RECORD

A. Length

1. Your height in inches ____________
2. Your height in centimeters ____________
3. Length of test tube in inches ____________
4. Length of test tube in centimeters ____________
5. cm/in. for your height readings ____________

   cm/in. for test tube readings ____________

   cm/in. from conversion table ____________

B. Weight

1. Unknown specimen ____________ g
2. Filter paper ____________ g
3. 250-mL beaker ____________ g

C. Volume

1. 250-mL beaker ________________ mL
2. Volume of water plus specimen ________________ mL

   Volume of water ________________ mL

   Volume of specimen ________________ mL

D. Density

1. Specimen

   Weight of specimen ________________

   Volume of specimen ________________

   Density of specimen ________________

2. Liquid

   Weight of 250-mL beaker and liquid ________________

   Weight of 250-mL beaker ________________

   Weight of liquid ________________

   Volume of liquid ________________

   Density of liquid ________________

Calculations

Name ______________________ Class ____________ Room ____________

EXERCISE 2
IDENTIFICATION OF A SUBSTANCE

INTRODUCTION

Properties are groups of characteristics that enable us to tell one substance from another. There are two main types of properties: physical and chemical. Physical properties are characteristics that usually do not involve a change from one substance into another. Such properties include color, solubility, density, odor, melting point, freezing point, and electrical conductivity. Determination of chemical properties usually results in a new substance being formed. Chemical properties include combustibility, acidic or basic character, and reaction with oxygen.

Physical properties can be determined objectively by means of scientific instruments. For a pure substance, physical properties remain constant at fixed temperatures and pressures.

Solubility is determined by measuring how much of a substance will dissolve in a liquid. It can be determined either quantitatively (for example, how many grams of a substance will dissolve in 100 mL of liquid?) or qualitatively (for example, does the substance completely dissolve, partially dissolve, or not dissolve?).

The boiling point depends upon the atmospheric pressure. When the vapor pressure of a liquid (the pressure of the gas formed by liquid evaporating) is equal to the atmospheric pressure, the liquid is said to boil. The temperature at which the vapor pressure of a liquid is equal to 1 atmosphere (atm) pressure is called the normal boiling point.

PRELAB STUDY QUIZ

1. What does boiling mean? How does it differ from evaporation?

2. How is boiling point of a liquid defined?

3. Give three properties that can be used to identify a substance.

4. A 50.0-g block of copper has a volume of 5.58 mL. What is the density of copper?

PROCEDURE

Obtain an unknown from your instructor. Record its number or letter. Enter all results on the Data Record.

A. Solubility

Add a couple of crystals of naphthalene to each of three test tubes, the first containing 2 mL of water, the second 2 mL of ethyl alcohol, and the third 2 mL of toluene. Stopper each test tube and shake briefly. Cloudiness indicates insolubility. Record whether the naphthalene is soluble or insoluble. Repeat using 2 drops of xylene and the same three solvents. Repeat using 2 drops of your unknown.

B. Density

Carefully weight your smallest beaker to the nearest 0.01 g. Record the weight. Pour about 20 mL of your unknown into a test tube or another beaker. Your instructor will demonstrate how to use a pipet and pipet bulb. (***Danger!*** *Do not pipet orally, as some of the unknowns are quite toxic if taken internally.)* Fill your pipet to the mark and transfer the unknown liquid to the previously weighed beaker. Weigh the beaker and its contents immediately. Record the volume of the pipet used. (This is the volume of the liquid transferred.) Calculate the weight of the unknown liquid and its density.

C. Boiling Point

Assemble the apparatus shown in Figure 2-1. The hose must reach the drain. Insert the thermometer into a cork stopper *carefully* to avoid breaking. Fill the test tube about one-fourth full with the unknown and add two boiling chips. The thermometer should be about 1 cm above the liquid level. Put the test tube into the middle of the water bath. Heat the water and watch for signs of boiling in the unknown. Initial bubbling may be due to dissolved gases escaping, so make sure the temperature remains constant for a couple of minutes after boiling begins. Record the boiling point. Using the solubility data, density, and boiling point of your unknown, determine its identity from Table 2-1.

Name ______________________ Class ______________ Room ____________

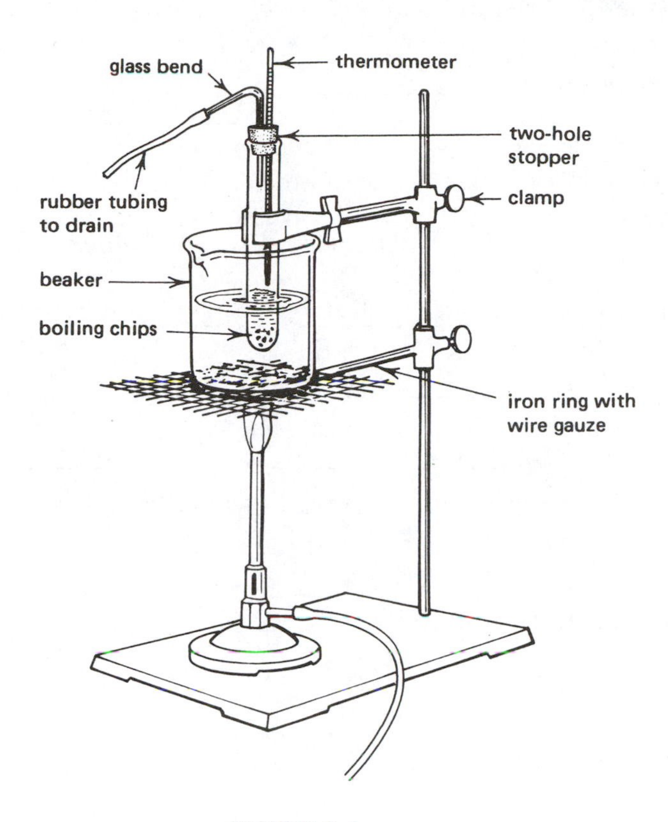

FIGURE 2-1

TABLE 2-1

PHYSICAL PROPERTIES OF PURE SUBSTANCES

Substance	Density (g/mL)	Boiling Point (°C)	Solubility* water	Solubility* benzene	Solubility* ethyl alcohol
Benzene	0.88	80	I	S	S
Chloroform	1.49	61	I	S	S
Ethyl acetate	0.90	77	S	S	S
Heptane	0.68	98	I	S	S
Isopropyl alcohol	0.79	83	S	S	S
Methyl (wood) alcohol	0.79	65	S	S	S
Methylene chloride	1.34	40.1	I	S	S
Propionaldehyde	0.81	48.8	S	I	S
Others (furnished by laboratory instructor)					
1. ______					
2. ______					
3. ______					

*S = soluble; I = insoluble

DATA RECORD

Unknown number or letter ____________

A. Solubility

	Water	*Toluene*	*Ethyl Alcohol*
Naphthalene	______	______	______
Xylene	______	______	______
Unknown	______	______	______

B. Density

Weight of beaker and unknown ______

Weight of beaker ______

Weight of unknown sample ______

Volume of unknown sample ______

Density of unknown $\left(\frac{\text{weight}}{\text{volume}}\right)$ ______

C. Boiling Point

Boiling point of unknown ____________

D. Identity of Unknown ____________

QUESTIONS

1. Explain the effect of atmospheric pressure on boiling point.

2. Why are size and weight by themselves not useful properties for the identification of an unknown?

3. Give an example of a liquid with a high density; a low density.

   What is the density of water?

4. Which of the properties you determined in this experiment do you consider to be the most useful in the identification of your unknown? Why?

5. Why are boiling chips useful when boiling a liquid?

Name ______________________ Class ____________ Room __________

EXERCISE 3
SEPARATION OF A MIXTURE

INTRODUCTION

From early times the separation and purification of substances has presented problems. Even today the separation and purification of mixtures is an important part of chemistry. Chemists must know the material they are working with if they are to correctly interpret the results of his experiments. For public safety, it is essential to know the purity of such things as medications, drugs, and foods.

Pure substances are either elements or compounds, each with a set of characteristic properties. Salt, sugar, and iron, for example, each have a set of properties that enables us to tell them apart. In Exercise 2 you were concerned with using various properties to identify an unknown. In this exercise you will be concerned with using different types of properties to separate the components of a mixture. You will be involved in three different purification procedures:

1. *Distillation.* Distillation is the separation of a liquid from a mixture by boiling it, condensing the vapors, and then collecting the condensed liquid. This process is used when a liquid mixture contains substances with large differences in boiling points.
2. *Extraction.* Extraction is the removal of a substance from a mixture by utilizing its greater solubility in a given liquid.
3. *Filtration.* Filtration is the separation of a solid from a liquid by passing the mixture through a porous material.

In this experiment you will separate a mixture containing sand, salt, water, and an oil-like compound. This experiment is somewhat similar to that presented to an environmentalist when an oil spillage occurs at sea. Although some of the separation procedures are different because of the differences in scale between the laboratory situation and an ocean spill, the basic techniques remain the same.

The basic separation scheme is outlined at the top of the following page.

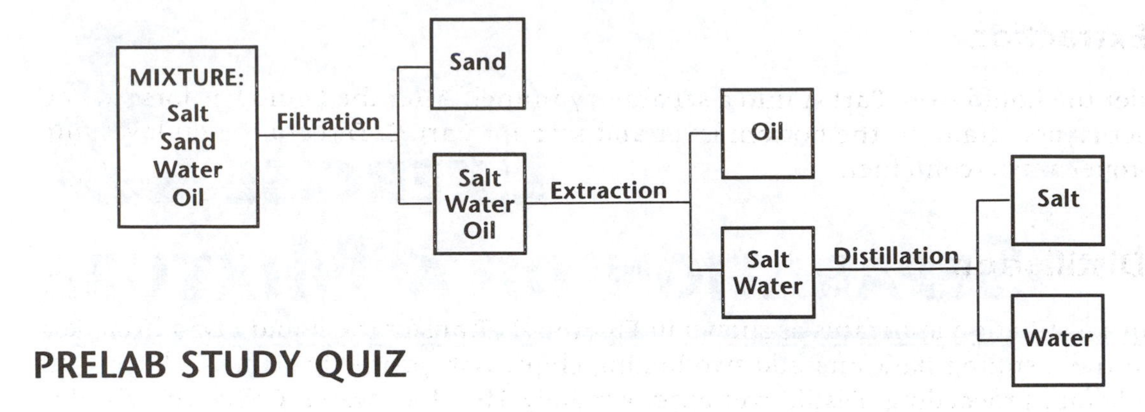

PRELAB STUDY QUIZ

1. How would you separate alcohol and water?

2. How would you separate sugar and sand?

3. There are two types of mixtures: homogeneous and heterogeneous. Homogeneous mixtures are uniform throughout; that is, they consist of only one single phase. Heterogeneous mixtures are not uniform throughout, and on visual examination the separate substances can be observed. Which type of mixture is sand/water? salt/water? oil/water?

4. Find a practical use for distillation, for extraction, for filtration.

SPECIAL HAZARD

Avoid spilling silver nitrate on your skin. Although the skin discoloration (black) is not a health hazard, the results are considered unsightly.

PROCEDURE

A. Filtration

Obtain your experimental mixture form the instructor or lab assistant. Fold a piece of filter paper as shown in Figure 12, page 9. Set up a filtration apparatus and filter your mixture as shown in Figure 13, page 10. Once the filtration is completed, remove the filtered liquid and save it for the next step.

Name ______________________ Class __________ Room __________

B. Extraction

Transfer the liquid from Part A into a separatory funnel. After the liquid has formed two distinct layers, drain off the bottom layer and save for Part C. Discard the oil layer into the proper waste container.

C. Distillation

Set up a distillation apparatus as shown in Figure 3-1. Transfer the liquid saved from Part B into the distilling flask and add two boiling chips. Ask your instructor to check your setup before proceeding. Distill over approximately 10 mL of water. Check the distilled water for chloride content by adding a drop of silver nitrate to it. A white, cloudy precipitate indicates the presence of chloride. After allowing the distilling flask to cool, check it for chloride content.

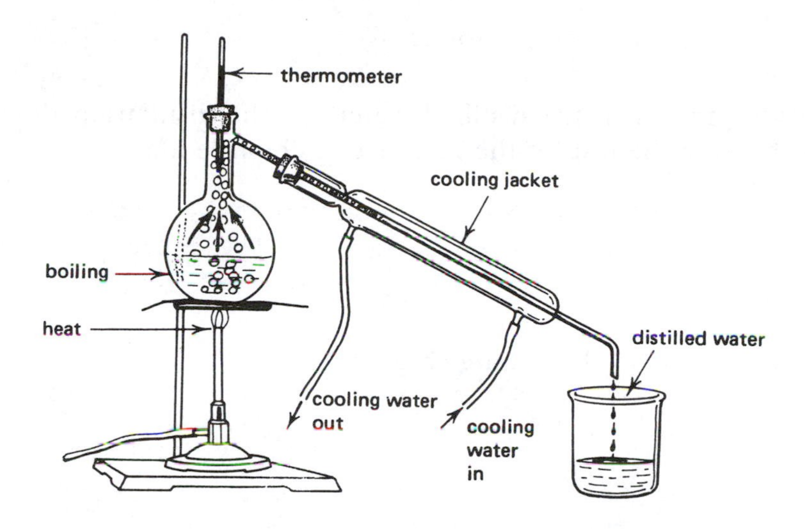

FIGURE 3-1

QUESTIONS

1. Which of these separation methods are used commercially to purify drinking water?

2. What are some medical uses for distilled water?

3. What methods are currently in use for cleaning up oil spills at sea?

4. In the separatory funnel, which is the oil layer? Why?

5. Is the recovered sand pure? Explain.

6. Is the recovered water pure? Explain.

7. Was chloride present in the distilled water? in the distillation flask? Write an equation showing the test for the presence of chloride ions.

8. What is the purpose of the boiling chips?

Name ______________________ **Class** __________ **Room** __________

EXERCISE 4
FORMULA OF A HYDRATE

INTRODUCTION

Some compounds, even though they are solids, contain a certain number of weakly held water molecules. These compounds are called hydrates, and the number of water molecules held by a particular compound is usually constant. The number of water molecules in the compound does vary the compound. For example, the formula $NiSO_4{\bullet}7H_2O$ means that there are 7 water molecules for every $NiSO_4$. Water molecules can usually be easily removed by heating.

$$NiSO_4{\bullet}7H_2O \xrightarrow{\Delta} NiSO_4 + 7H_2O$$

$NiSO_4{\bullet}7H_2O$		$NiSO_4$	+	$7H_2O$
mol. mass 281		mol. mass 155	+	mol. mass 18
281 g		155 g		126 g

Examples of other hydrates include $CuSO_4{\bullet}5H_2O$, $NaBr_2{\bullet}2H_2O$ and $Na_2CO_3{\bullet}10H_2O$. In this experiment you will determine the number of moles of water that combine with 1 mole of barium chloride in the hydrate $BaCL_2{\bullet}XH_2O$. As with the above example, when a hydrate is heated, water is lost.

$$BaCl_2{\bullet}XH_2O \xrightarrow{\Delta} BaCl_2 + XH_2O$$

The weight of water is the difference between the weight of barium chloride before and after heating. It is a simple calculation to determine X if the weight of $BaCl_2$ and the weight of H_2O are known.

Calculations

$$\text{Number of moles } BaCl_2 = \frac{\text{wt. of } BaCl_2 \text{ (after heating)}}{208 \text{ g/mole}}$$

$$\text{Number of moles } H_2O = \frac{\text{wt. of } H_2O \text{ (wt. of } BaCl_2{\bullet}XH_2O \text{ before heating} - \text{wt. of } BaCl_2 \text{ after heating)}}{18 \text{ g/mole}}$$

$$\text{Value of X} = \frac{\text{number of moles } H_2O}{\text{number of moles } BaCl_2}$$

PRELAB STUDY QUIZ

1. What is a hydrate?
2. Give two examples of hydrates.
3. What hydrate is used in this experiment?
4. How will you determine the amount of water in the hydrate?

PROCEDURE

1. Obtain a crucible and cover. Clean and dry them.
2. Weigh the crucible and cover to nearest 0.01 g. Record the mass.
3. Add about 3 g of $BaCl_2{\bullet}XH_2O$ to the crucible, cover, and weigh the crucible, cover, and contents to the nearest 0.01 g. Record the weight.
4. Place the crucible on a clay triangle on a ring stand. Leave the cover slightly open to allow water to escape. Heat the crucible gently with the burner for 4–5 minutes. Heat more strongly for another 4–5 minutes. Place the burner under the crucible and heat with the blue of the flame for about 10 minutes.
5. Turn off the flame, cover the crucible with its cover (Use tongs as cover is very hot), and allow to cool to room temperature (about 10 minutes). Weigh the crucible, cover, and contents. Record the mass.
6. To insure that all the water has been removed, put the crucible, cover and contents back on the clay triangle and reheat with blue flame for 5 minutes. Allow crucible, cover, and contents to cool for 10 minutes. Reweigh. This weight should be very close to weight obtained in step 5. If it isn't, repeat heating procedure.
7. Using your data in calculation example, calculate X for $BaCl_2{\bullet}XH_2O$.

DATA RECORD

Mass of empty dry crucible and cover ____________________

Mass of crucible, cover, and $BaCl_2{\bullet}XH_2O$. ____________________

Name ______________________ Class ____________ Room ____________

Mass of crucible, cover, and $BaCl_2$ after first heating ______________________

Mass of crucible, cover, and $BaCl_2$ after second heating ______________________

Mass of water lost ______________________

Mass of anhydrous $BaCl_2$ (anhydrous = without H_2O) ______________________

Calculate number of moles water lost. Show your work.

Calculate number of moles anhydrous $BaCl_2$. Show work.

Calculate value of X in $BaCl_2 \cdot XH_2O$. Show work.

Name ______________________ Class __________ Room __________

EXERCISE 5
EMPIRICAL FORMULA OF A COMPOUND

INTRODUCTION

The two different kinds of formulas that you will encounter in chemistry are empirical formulas, which give the simplest possible ratio of numbers of atoms in a compound, and molecular formulas, which give the actual number of atoms in a compound. For example, the molecular formula for hydrogen peroxide is H_2O_2. In a molecule of hydrogen peroxide two atoms of hydrogen are bonded to two atoms of oxygen. The empirical formula of hydrogen peroxide, which is the simplest ration of different kinds of atoms, is HO.

To determine the empirical formula you need to know the stoichiometric data. This is the information about the relative masses of different elements combined in the compound. In this experiment you will determine the empirical formula of a compound formed from zinc and chlorine.

PRELAB STUDY QUIZ

1. Explain the difference between empirical formula and molecular formula.

2. What is the molecular formula for hydrogen peroxide? The empirical formula?

3. What elements does the compound you are making contain?

SPECIAL HAZARD

Hydrochloric acid can cause severe skin and eye burns. If you spill any on you, wash the affected area with a large amount of water and notify your instructor.

PROCEDURE

1. Clean and dry the evaporating dish.
2. Place the dish on the wire gauze over the Bunsen burner. Heat, gently at first, then more strongly, for about 5 minutes.
3. Allow the dish to cool. Weigh and record its mass to 0.01 g.
4. Add about 0.5 g of zinc to the evaporating dish. Weigh and record the mass of the dish and the zinc.
5. IN THE HOOD, slowly add 15 mL of 6 M HCl to the zinc. Stir after the vigorous reaction has stopped. The wet hydrogen gas produced must be kept in the hood since it is flammable.
6. If any solid zinc remains, add an additional 5 mL of 6 M HCl (IN THE HOOD) until the zinc has completely reacted.
7. Place the evaporating dish on a hot plate IN THE HOOD. Heat until the liquid has evaporated and the dry solid remains. DO NOT HEAT TOO LONG, or the zinc chloride will melt.
8. Allow the dish and its contents to cool. Weigh and record the mass.
9. Gently heat the dish and contents again. Cool and reweigh. If the mass is different from that obtained in step 9 by more than 0.02 g, repeat this process until "constant" mass is obtained.
10. Repeat the entire experiment with a second sample of zinc. This may be started while the first dish is heating on the hot plate.

Sample Calculation

0.45 g of tin (Sn) combined with 0.12 g of oxygen. Determine the empirical formula of the oxide of tin formed.

Name ______________________ Class ____________ Room ____________

1. Convert grams of each element to moles.

$$\text{moles of Sn} = 0.45 \text{ g Sn} \times \frac{1 \text{ mole Sn}}{119 \text{ g Sn}} = 0.0038 \text{ moles Sn}$$

$$\text{moles of O} = 0.12 \text{ g O} \times \frac{1 \text{ mole O}}{16.0 \text{ g O}} = 0.0075 \text{ moles O}$$

2. Determine the whole number ratio of moles and atoms.

$$\frac{0.0038 \text{ mole Sn}}{0.0038} = 1 \text{ mole Sn}$$

$$\frac{0.0075 \text{ mole O}}{0.0038} = 1.97 = 2 \text{ moles O}$$

3. Formula is SnO_2

DATA RECORD

	Sample 1	*Sample 2*
Mass of evaporating dish	____________	____________
Mass of evaporating dish and zinc	____________	____________
Mass of evaporating dish and zinc chloride (1st heating)	____________	____________
Mass of evaporating dish and zinc chloride (2nd heating)	____________	____________

Calculations

Show your methods of calculation.

Mass of zinc	____________	____________
Mass of zinc chloride	____________	____________
Mass of chlorine	____________	____________
Formula of the zinc chloride (based on your calculations)	____________	____________

Name ______________________ Class ____________ Room ____________

Exercise 6
Chemical Reaction of Copper

INTRODUCTION

Most chemical syntheses involve separation and purification of the desired product from unwanted side products. Some methods of separation you have used already. This experiment is designed as a quantitative evaluation of your individual laboratory skills in carrying out some of these operations. At the same time you will become more acquainted with two fundamental types of chemical reactions—redox reactions and metathesis reactions. By means of these reactions, you will carry out several chemical transformations involving copper, and you will finally recover the copper sample with maximum efficiency. The chemical reactions involved are the following:

(1) $Cu(s) + 4\ HNO_3(aq) \rightarrow Cu(NO_3)_2(aq) + 2\ NO_2(g) + 2\ H_2O$ REDOX

(2) $Cu(NO_3)_2(aq) + 2\ NaOH(aq) \rightarrow Cu(OH)_2(s) + 2\ NaNO_3(aq)$ METATHESIS

(3) $Cu(OH)_2(s) \rightarrow CuO(s) + H_2O$ DEHYDRATION

(4) $CuO(s) + H_2SO_4(aq) \rightarrow CuSO_4(aq) + H_2O$ METATHESIS

(5) $CuSO_4(aq) + Zn(s) \rightarrow ZnSO_4(aq) + Cu(s)$ REDOX

Each of the reactions proceeds to completion. Metathesis reactions proceed to completion whenever one of the components is removed from the solution, such as in the formation of a gas or an insoluble precipitate. This is the case for reactions (1), (2), and (3), where in reaction (1) a gas, and in (2) and (3) an insoluble precipitate, are formed. Reaction (5) proceeds to completion because zinc has a lower ionization energy or oxidation potential than copper.

The object in this experiment is to recover all of the copper you began with in analytically pure form. This is the test of your laboratory skills.

The percent yield of the copper can be expressed as the ratio of that obtained to that expected, multiplied by 100:

$$\% \text{ yield} = \frac{\text{experimental yield Cu}}{\text{calculated yield Cu}} \times 100$$

PRELAB STUDY QUIZ

1. What acid is used to dissolve the copper?

2. Write the equation for the dehydration reaction.

3. In what form is the copper recovered?

4. What metal is used to reduce the copper in copper sulfate?

SPECIAL HAZARD

Nitrogen dioxide is a poisonous red-brown gas. The reaction producing nitrogen dioxide should only be done in the hood. Nitric and sulfuric acids are very corrosive acids that will cause severe skin and eye burns. Wash any affected area with large amounts of water and notify the instructor.

PROCEDURE

1. Determine the mass of approximately 0.50 g of copper wire to the nearest 0.01 g and place it in a 250 mL beaker.

2. IN THE HOOD, add 4–5 mL of concentrated nitric acid, HNO_3, to the beaker. On the data page, describe the reaction as to color change, evolution of gas, and change in temperature (exothermic or endothermic).

3. After the reaction is complete (all the Cu has dissolved and no more gas is given off), add 100 mL of deionized water and stir.

4. Add 30 mL of 3.0 M NaOH to the solution in the beaker and describe the reaction on the data page.

5. Add two or three boiling chips and *carefully* heat the beaker—while stirring with a stirring rod—until the solution *just starts to boil.* Describe the reaction on the data page.

6. Allow the black CuO to settle; then decant the supernatant liquid. Add about 200 mL of *very hot* deionized water, stir, and allow the CuO to settle. Decant the supernatant liquid again.

7. Add 15 mL of 6.0 M H_2SO_4 to the solid CuO. Remove the boiling chips. Describe the reaction on the data page.

8. IN THE HOOD, add 2.0 g of zinc metal all at once to the beaker and stir until the supernatant liquid is colorless. Describe the reaction on the data page.

9. When gas evolution becomes *very* slow, heat the solution gently on the steam bath in the hood (but DO NOT BOIL), then allow it to cool.

10. Weigh an evaporating dish. When gas evolution has stopped, decant the solution and transfer the copper precipitate to the weighed evaporating dish.

11. Wash the precipitated copper with about 5 mL of deionized water; allow it to settle; decant the solution; and repeat the process of washing, settling, and decanting. (OPTION: Wash with acetone.)

12. Dry the product in the evaporating dish on a steam bath at least 5 minutes. Watch carefully and remove as soon as it looks dry.

13. Weigh the cooled evaporating dish and dry copper.

DATA RECORD

Mass of copper wire ____________

Mass of evaporating dish ____________

Mass of evaporating dish and copper ____________

Mass of recovered copper ____________

Percent yield (show calculations) ____________

Observations

QUESTIONS

1. Describe the reaction in step 2 as to color change, evolution of gas, and temperature change.
2. What color is the copper nitrate solution?
3. What color is nitrogen dioxide gas?
4. Describe what forms in step 4.
5. What color is copper (II) hydroxide?
6. What gas is produced in step 8?
7. Write an equation for the production of the gas in step 8.

Name ______________________ Class ____________ Room ____________

EXERCISE 7
RADIOACTIVITY

INTRODUCTION

A radioactive element is an element that spontaneously gives off radiation and, by so doing, usually changes into another element. Some radioactive elements occur naturally, whereas others are manmade. Each radioactive element emits radiation that is characteristic of that element, both in type and energy. The three principal types of radioactive particles, or radiation, of interest in this experiment are alpha, beta, and gamma. Table 7-1 shows their properties.

TABLE 7-1

Name	Symbol	Charge	Mass (amu)	Identity
Alpha	α	+2	4	Helium nucleus
Beta	β	-1	≈ 0	High-speed electron
Gamma	γ	0	0	Electromagnetic radiation

The decay of radium by the loss of an alpha particle to form the element radon can be summarized as

$$^{226}_{88}\text{Ra} \rightarrow {}^{222}_{86}\text{Rn} + \alpha\ ({}^{4}_{2}\text{He})$$

Loss of an alpha particle results in a decrease of 2 in atomic number and a decrease of 4 in mass number.

Loss of a beta particle involves the splitting of a neutron into a proton and an electron. The discharge of the electron (beta particle) from the nucleus increases the positive charge of that nucleus, thereby increasing the atomic number by 1.

$$^{32}_{15}P \rightarrow {}^{32}_{16}S + \beta\,({}^{0}_{-1}e)$$

Emission of gamma radiation often accompanies alpha or beta emission, but production of gamma radiation does not result in the formation of a new element. Gamma radiation is merely a form of radiation energy.

Radioactive elements vary in their stability or rate of radioactive decay. Relative stabilities of radioactive elements are measured in terms of half-lives ($t_{1/2}$), that is, the time required for half of the radioactive atoms to decay.

There are many medical uses for radioactive elements, including cancer treatment, body scans, and metabolism studies. Even with safeguards against loss or dispersion of radioactive materials, hazards may exist. The following set of rules is applicable to both laboratory and hospital:

1. Do not touch any radioactive source directly with the hands. Handle such sources with forceps.
2. Report all wounds and spills immediately to the instructor.
3. Before leaving, wash your hands and check for radiation. Radioactivity remaining after thorough washing should be reported.

In addition:

4. Handling of or exposure to radioactive materials under ordinary conditions does *not* make a person radioactive. Radioactivity is dangerous only because of the destructive effect it has on tissue.
5. All persons are constantly exposed to unavoidable radioactivity, called background radiation. This arises from such sources as cosmic rays and from natural and manmade radiation in the air, soil, and water.

There are two ways to protect yourself against radiation: (1) Increasing the distance to the source of radiation. (The amount of radiation you receive decreases as the distance to the source increases.) (2) Placing a shield between yourself and the source of radiation. (Various materials placed in the path of radiation will reduce the intensity of that radiation.)

PRELAB STUDY QUIZ

1. By what two ways can you protect yourself from radiation?

2. When a radioactive isotope of strontium decays, it becomes yttrium. What type of radiation was emitted?

Name ______________________ Class ____________ Room ____________

3. Name three sources of background.

4. Name three medical uses for radioactive isotopes.

PROCEDURE

1. The instructor will show you how to use the Geiger counter. Record all your results on the Data Record.
2. Determine the background count per minute.
3. Place the radioactive source or sources at 5 cm, 10 cm, 20 cm, and 40 cm from the unshielded tube, as shown in Figure 7-1. Record the counts per minute at each distance.
4. Determine the counts per minute at 20 cm, using the following as shields: cardboard, lead, wood, brick (pottery), clear glass, cobalt glass, aluminum, and cotton of approximately equal thickness.
5. Determine the net count per minute for each of the preceding measurements by subtracting the background count from each of the readings.

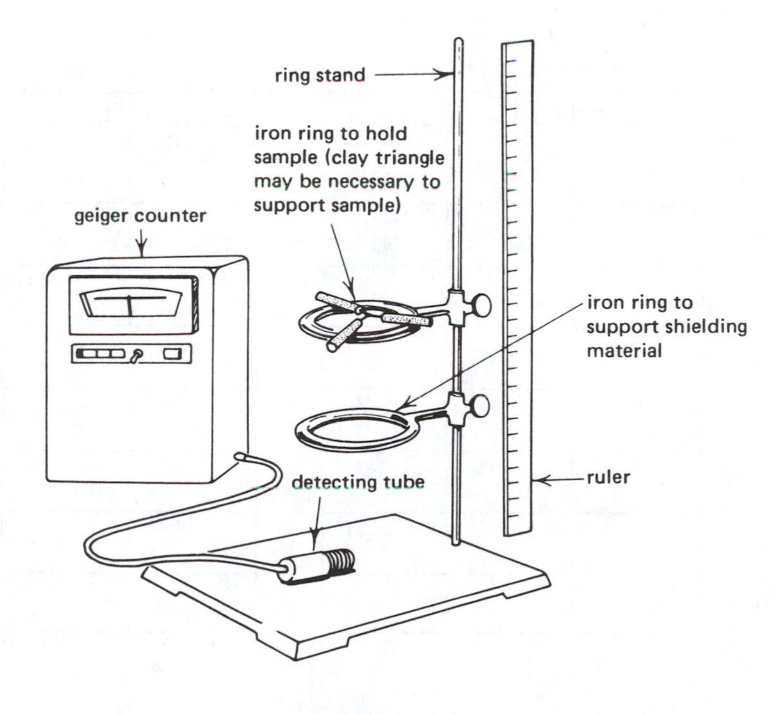

FIGURE 7-1

DATA RECORD

Background count ____________

Radioactive source (1) ____________ (2) ____________ (3) ____________

Particle emitted (1) ____________ (2) ____________ (3) ____________

For air:

	Measured Counts/Min			Net Count/Min		
	Source 1	Source 2	Source 3	Source 1	Source 2	Source 3
5 cm						
10 cm						
20 cm						
40 cm						

Shielded radiation counts at 20 cm:

	Measured Counts/Min			Net Count/Min		
	Source 1	Source 2	Source 3	Source 1	Source 2	Source 3
Cardboard						
Lead						
Wood						
Brick (pottery)						
Clear glass						
Cobalt glass						
Aluminum						
Cotton						

Name ______________________ Class ____________ Room ____________

QUESTIONS

1. What effect does doubling the distance have on the net count?

2. Which materials are effective shields? Which are not?

3. Which type of radiation is the most penetrating?

CASE HISTORY: HYPOTHYROID CONDITION

Mrs. Winters, a 38-year-old patient, complained of fatigue as well as dizziness on rising suddenly. She had recently gained weight and had low blood pressure.

Palpation of the thyroid gland showed a small nodule. Blood was drawn for T_4 and T_3 uptake. A scan was ordered. The results of the scan showed a cold nodule in the thyroid gland.

CLINICAL ASPECTS

MICHAEL REESE HOSPITAL AND MEDICAL CENTER

015284

UNIT RECORD NUMBER | Y | MO | DAY | YR | H U | 1 | 2 | 3 | 4 | 5 | 6 | DEPT. CODE 603

☒ PRIVATE OUTPATIENT ☐ EMERGENCY ROOM ☐ MANDEL CLINIC

DIAGNOSIS — TOTAL CHARGE — MRS. SPRING WINTERS

SPECIMEN — DATE AND TIME COLLECTED — DATE EXAMINED

RESIDENT OR INTERN B. Cohen — REQUISITION PREPARED BY — TECH.

CHART

Test	Normal		Test	Normal		Test	Normal		Test	Normal	
SGOT ☐ 137	<35 U.		Total Protein ☐ 144	6.3-7.2 gm %		Electrophoresis ☐ 149	SEE SEPARATE REPORT		CHOLESTEROL ☒ 153 TOTAL	160-220 mg %	250
SGPT 138 ☐	<35 U.		☐ ALBUMIN AND	3.8-4.8 gm %		BSP 150 ☐ 5 min.			CHOLESTEROL 154 ☐ FREE	mg %	
LDH ☐ 139	200-500 U.		145 A/G RATIO	1.5-2.5		45 min.			% FREE	22-28 % of Total	
ALK. P'TASE 140 ☐	Ad. 2-4 U. Ch. 4-8 U.		INSTRUCTIONS						CEPH. FLOCC. ☐ 155	Neg.	
ACID P'TASE ☐ 141 TOTAL	<2.0 U.								THYMOL TURBIDITY 156 ☐	<2.7	
PROSTATIC	<1.0 U.								THYMOL FLOCC. ☐ 157	Neg.	
URIC ACID 142 ☐	3.0-6.0 mg %		CPK ☐ 147	28U-44U		Triglycerides ☐ 151	30-130 mg %		ALDOLASE 158 ☐	M <33 U. F <19 U.	
ASAL ☐ 143	<5 U.		TOTAL LIPIDS 148 ☐	Ch. 200-400 Ad. 400-800		Phospholipids 152 ☐	175-350 mg %		XYLOSE ☐ 159 (Blood)	SEE SEPARATE REPORT	

MICHAEL REESE MEDICAL CENTER

265180

UNIT RECORD NUMBER | V | B | MO | DAY | YR | N | U | 1 | 2 | 3 | 4 | 5 | 6

DEPT. CODE 613 ☐ MRHP ☐ PRIVATE OUTPATIENT ☐ EMERGENCY ROOM ☐ MANDEL CLINIC

DIAGNOSIS

TOTAL CHARGE

SPECIMEN: BLOOD

DATE AND TIME COLLECTED

DATE EXAMINED

RESIDENT OR INTERN

REQUISITION PREPARED BY

TECH.

15cc OF CLOTTED BLOOD REQUIRED FOR STUDIES LISTED BELOW

Test	Normal	Result
PLASMA PBI ☒ 111	3.5-8.0 mcg %	3.0
THIS STUDY CANNOT BE DONE IF THE PATIENT IS ON EXOGENOUS IODINE OR HAS HAD AN IVP, ARTERIOGRAM, OR GALL BLADDER X-RAY RECENTLY.		
T3 RESIN 112 ☒	24-38 %	19
ANTI-THYROGLOBULIN ☐ 113 ANTIBODY	NEG.	
SERUM THYROXINE (T4) 114 ☒	5-13.5 mcg %	4.1

15cc EDTA (LAVENDER TOP TUBE) BLOOD REQUIRED FOR:

15cc OF CLOTTED BLOOD REQUIRED FOR STUDIES LISTED IN THIS SECTION.

Test	Normal	Result
RISA PLASMA VOLUME ☐ 117		
BLOOD VOLUME		
RED CELL MASS		
TRIIODOTHYRONINE ☒ 151	50-160 Ngm %	43

CHART

Test	Normal
DIGOXIN ASSAY ☐ 152	ON .50 mg./DAY 0.9-2.4
153 ☐ CEA ANTIGEN	
RENIN RIA ☐ 154	Low Salt: 3.00-6.00 ng/ml/hr. Norm. Salt: .75-2.75 ng/ml/hr.
ALDOSTERONE RIA 155 ☐	2-26 µg/24 hr.
FERROKINETICS ☐ 156	60-120 min.
RED CELL SURVIVAL 157 ☐ SEQUESTRATION	26-34 DAYS
T B G ☐ 158	15-23 µg %
666 ☐	CHARGE $

Reprinted by permission of Michael Reese Hospital and Medical Center, Chicago.

1. Look at the lab reports on pages 43 and 44. What tests were ordered?

   Which results are within the normal range? Which are not?

2. Which isotopes are used in the diagnosis and treatment of thyroid conditions?

3. What is a cretin? What characteristics do they have?

4. What are the symptoms of myxedema? How may it be treated?

5. What is a "simple goiter"? How may it be treated?

6. What are the symptoms of Graves' disease?

Name ______________________ Class ____________ Room ____________

7. What are the symptoms of Plummer's disease?

8. How are the preceding diseases treated?

9. List some antithyroid drugs and state how they function.

10. What is exophthalmos? How may it be treated?

11. A lack of calcitonin will cause what clinical symptoms?:

12. Why was a cholesterol test ordered along with the thyroid function test?

BASIC SCIENCE ASPECTS

1. Where is the thyroid gland located?

2. What hormones does the thyroid gland produce?

3. What is a PBI? a BEI? What test now generally replaces these?

4. What might cause hypothyroidism?

5. What is myxedema?

6. What might cause hyperthyroidism?

7. How is TSH related to thyroid function? Where is it produced?

8. What controls the production of TSH? How is the production regulated?

9. How is thyroid function related to metabolic rate?

10. What is T_3? T_4? Where are they produced? What is their function?

Name ______________________ Class ______________ Room ______________

EXERCISE 8
GASES

INTRODUCTION

Gases are of vital importance to us in many ways. They form our atmosphere (largely nitrogen and oxygen); support life (oxygen); heat our homes (methane); serve as important raw materials (nitrogen and hydrogen); and are the principal air pollutants (oxides of sulfur, nitrogen, and carbon, as well as ozone). In this experiment you will prepare a variety of gases and study some of their chemical properties. You will also study the rates of diffusion of two gases. At the same temperature, the rate of diffusion depends only on the molecular weight of the gas involved. The lighter the gas, the faster it diffuses.

PRELAB STUDY QUIZ

1. The relative rates of diffusion of gases depend upon what factors?

2. What gas is used for heating purposes?

3. Name some gases that contribute to air pollution. What is the source of each?

4. Which should diffuse faster, hydrogen or oxygen? oxygen or carbon dioxide? ammonia or methane?

SPECIAL HAZARD

Hydrochloric acid can cause severe skin and eye burns. If you spill any on you, wash the affected area with a large amount of water and notify your instructor.

PROCEDURE

The following procedures are to be carried out in pairs. All results are to be entered in the Data Record.

A. Preparation and Reactions of Oxygen

1. *Preparation.* Fill two wide mouth bottles with water, and invert them into a water trough half filled with water. Place 25 mL of 3% hydrogen peroxide into the oxygen-generating tube, add 1 mL 0.1 M iron (III) chloride, and connect as shown in Figure 8-1. Gently warm the generating tube, using a low flame. Stop heating as soon as the production of oxygen begins, that is, as soon as gas begins bubbling out. (***Caution:*** *Do not boil.*) Once oxygen production begins, insert the rubber tubing into one of the water-filled test tubes. When this tube is filled; fill the second tube. Stop further production of oxygen by rinsing the generating tube. Stopper the tubes and place upright in a test tube rack.

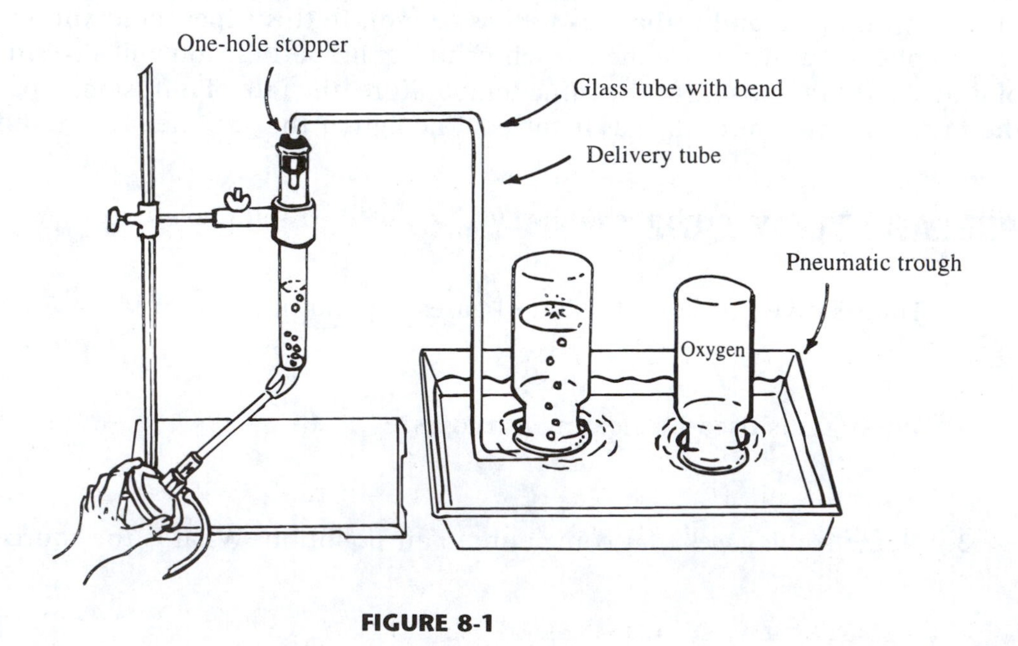

FIGURE 8-1

2. *Reaction with Carbon.* Heat a small piece of charcoal in a deflagrating spoon, or a small spoon made of aluminum, until it glows. Remove the stopper from one of the tubes of oxygen and quickly immerse the spoon with the *glowing* char-

coal into it. After the reaction ceases, remove the spoon, add 5 mL of water to the test tube, stopper, and shake. Test for the acidic or basic character of the liquid by using red and blue litmus paper.

3. *Reaction with Sulfur.* Repeat the procedure in paragraph 2 using sulfur in place of carbon. (***Note:*** *This part must be performed under the hood since the fumes are toxic.*)

B. Preparation of Hydrogen

Place a couple of pieces of mossy zinc in a test tube and add 1 to 2 mL of dilute HCl. The gas being generated is hydrogen. Invert a dry, empty test tube over the mouth of the first tube and allow it to fill with the hydrogen gas for a couple of minutes. Place your thumb over this tube and, while holding the inverted test tube at arm's length, remove your thumb and bring a lighted splint to the mouth of the test tube. A sharp "pop" (slight explosion) should occur if hydrogen is present. (***Caution:*** *Do not point the test tube at anyone when igniting the hydrogen, and do not hold the tube too close to you.*)

Record your results, being sure to note what was formed on the insides of the test tube. To stop the production of hydrogen, rinse the generating tube. Do not pour the unreacted zinc into the sink.

C. Preparation of Carbon Dioxide

You will need the rubber stopper with the glass bend and tubing shown in Figure 8-1, Part A. Place three or four marble chips in a test tube. Half fill a second test tube with limewater and insert the rubber tubing beneath the surface of the liquid. Pour 2 to 3 mL of dilute HCl into the test tube containing the marble chips. Place the stopper apparatus from Part A into this tube and allow the carbon dioxide being generated to pass through the limewater for 10 to 15 minutes. Record any changes you observe. Can you account for those changes?

D. Preparation of Chlorine

Place about 0.5 g of calcium hypochlorite in a test tube. (***Caution:*** *This part of the experiment must be performed under the hood because chlorine gas is highly toxic.*) Cautiously add 10 mL of 6 M H_2SO_4 to the test tube. Record your observations. Place a burning splint into the gas. Rinse the test tube in the sink under the hood.

E. Preparation of Oxides of Nitrogen

Nitric oxide (NO) and nitrogen dioxide (NO_2) are poisonous and must be prepared under the hood. Place about 0.5 g of $NaNO_2$ into a test tube and add 1 mL of dilute HCl. Place a burning splint into the gas being generated and record your observations. Rinse the test tube in the hood sink.

F. Diffusion

Assemble the apparatus shown in Figure 8-2. The glass tube should be about 50 cm long and about 1 cm in diameter. The tube should be clamped into position so that it is perfectly level. Dry the inside of the tube by gently warming it. Place a small piece of cotton at each end of the tube and insert one-hole stoppers or corks at each end. Add about 5 drops of concentrated ammonium hydroxide to the cotton at one end by inserting the tip of a medicine dropper through the hole of the stopper. Simultaneously add 5 drops of concentrated hydrochloric acid to the cotton at the other end. Note the time. Watch for the formation of a white deposit of ammonium chloride inside the tube and note the time at which it forms. Measure the distance in centimeters that each gas diffused from the end of the tube to the ammonium chloride deposit. Which gas diffused faster, ammonia or hydrogen chloride? (The ammonia comes from the ammonium hydroxide and the hydrogen chloride from the hydrochloric acid.)

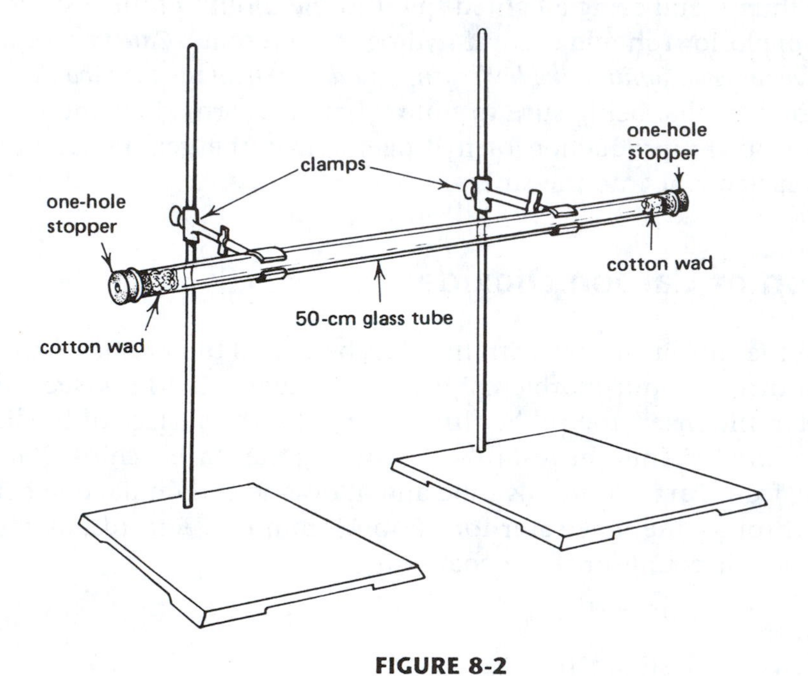

FIGURE 8-2

DATA RECORD

A. Preparation and Reactions of Oxygen

What gas was produced in Part 2? Was a water solution of this gas acidic, basic, or neutral?

What gas was produced in Part 3? Was a water solution of this gas acidic, basic, or neutral?

Name ______________________ Class ____________ Room ____________

B. Preparation of Hydrogen

Describe the results when the lighted splint was brought near the hydrogen.

What was formed inside the test tube?

C. Preparation of Carbon Dioxide

Observations of reaction of carbon dioxide with limewater.

Explain these changes.

D. Preparation of Chlorine

Observation of preparation of chlorine and reaction with flame.

E. Preparation of Oxides of Nitrogen

Observation of preparation and reaction with flame.

F. Diffusion

Time at start of diffusion ______________

Time at end of diffusion ______________

Time of diffusion (seconds) ______________

Distance travelled by NH_3 gas ____________ cm

Distance travelled by HCl gas ____________ cm

Rate of diffusion of NH_3 $\left(\frac{\text{distance}}{\text{time}}\right)$ ____________

Rate of diffusion of HCl $\left(\frac{\text{distance}}{\text{time}}\right)$ ____________

Which gas diffused faster? ____________

How many times faster? ____________

Name ______________________ Class ____________ Room ____________

Exercise 9
Percent Potassium Chlorate in a Mixture

INTRODUCTION

$KClO_3$ decomposes upon heating according to the equation:

$$2\ KClO_3(s) \xrightarrow{MnO_2} 2\ KCl(s) + 3\ O_2(g)$$

In this experiment you will be given a mixture of $KClO_3$ and KCl, with MnO_2 as a catalyst. After the mixture is weighed, it will be heated and the volume of oxygen gas produced will be measured under the conditions of temperature and pressure in the laboratory. The ideal gas equation ($PV = nRT$) can then be used to calculate the number of moles of oxygen, and using stoichiometry, the mass and mass percent of $KClO_3$ in the original mixture can be found.

PRELAB STUDY QUIZ

1. What gas is being produced?

2. What is the source of this gas?

3. What is the function of the manganese dioxide?

4. Write the ideal gas equation.

5. Find the value for *R* in your text.

6. What are the units for *P, V, T,* and *n*?

PROCEDURE

1. Measure the mass of the dry, empty test tube to the nearest 0.01 g on the beam balance.

2. Add about 0.6 g of the unknown sample mixture to the test tube and measure the mass of the test tube and contents.

3. Set up the apparatus as shown in Figure 9-1. the instructor MUST check your system BEFORE you begin heating. Pinch clamp must be open before beginning.

4. Heat the test tube gently at first. As the rate of oxygen production begins to slow, heat it more intensely until no more water is seen flowing into the beaker.

5. Stop heating and allow the system to cool to room temperature. Be certain that the glass tube is below the water level in the beaker.

6. When the system is cool equalize the pressure in the beaker and flask by lifting the beaker so the water levels are the same. Clamp the rubber tubing.

7. Use a graduated cylinder to measure the volume of water in the beaker.

8. Measure the temperature of the water in the flask.

9. Record the atmospheric pressure.

10. Repeat the experiment with a second sample.

Name ______________________ Class ____________ Room ____________

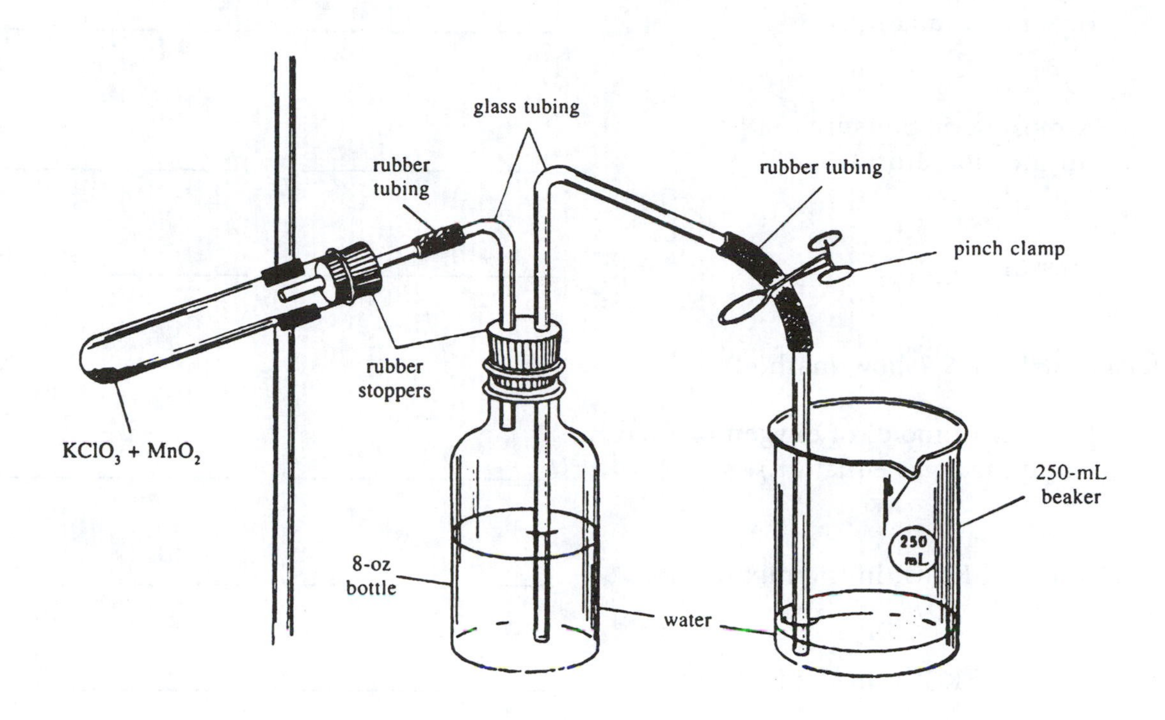

FIGURE 9-1

DATA RECORD

Instructor's Initials ____________

	Sample 1	*Sample 2*
Mass of empty test tube	________	________
Mass of test tube and sample	________	________
Mass of sample	________	________
Volume of water in beaker (volume of O_2 produced)	________	________
Temperature of water	________	________

Vapor pressure of water
(from text table) ______________ ______________

Atmospheric pressure
(in the laboratory) ______________ ______________

Pressure of O_2 ______________ ______________

Calculations (show method)

Number of moles of oxygen evolved
(from ideal gas equation) ______________ ______________

Moles of $KClO_3$ in the mixture ______________ ______________

Grams of $KClO_3$ in the mixture ______________ ______________

Percent $KClO_3$ in the mixture ______________ ______________

Name ______________________ Class __________ Room __________

EXERCISE 10
MOLAR MASS OF A VAPOR

INTRODUCTION

The ideal gas law ($PV = nRT$) enables you to calculate the molar mass of a gas if you know the mass of the gas and its P, V, and T. Since $n = g/molar\ mass$, it may be substituted in the ideal gas law to give $PV = gRT/molar\ mass$. If you rearrange this to solve for molar mass, you get

$$\text{molar mass} = \text{gRT}/PV$$

You will determine g, P, V, and T. Using the value of R in your textbook (with the proper units), the molar mass of the vapor can be calculated.

PRELAB STUDY QUIZ

1. What are you determining in this experiment?

2. What four measurements are necessary to calculate the molar mass of the vapor?

3. A 256-mL flask, at 373 K and 750 mm Hg, contains 0.80 g of a gas. Calculate the molar mass of the gas.

PROCEDURE

1. Place a small square of aluminum foil over the top of a clean, dry 125-mL Erlenmeyer flask. Fold the foil around the neck of the flask and secure it with a rubber band. Make a very small hole in the middle of the foil with a pin. (See Figure 10-1.)
2. Determine the mass of the flask and cap to the nearest 0.01 g. (0.0001g if you use an analytical balance.)
3. Remove the cap and place 6.0 mL of the unknown liquid in the flask and replace the cap.
4. Fill the 600-mL beaker with enough water so that it can surround the flask when it is immersed in the beaker.
5. Heat the water to boiling.
6. Clamp the flask inside the beaker of boiling water. Watch the liquid in the flask.
7. When all the liquid in the flask has vaporized, wait 30 seconds, then remove the flask from the water bath and set it aside to cool. While the water is still boiling, record the temperature of the water and note the barometric pressure.

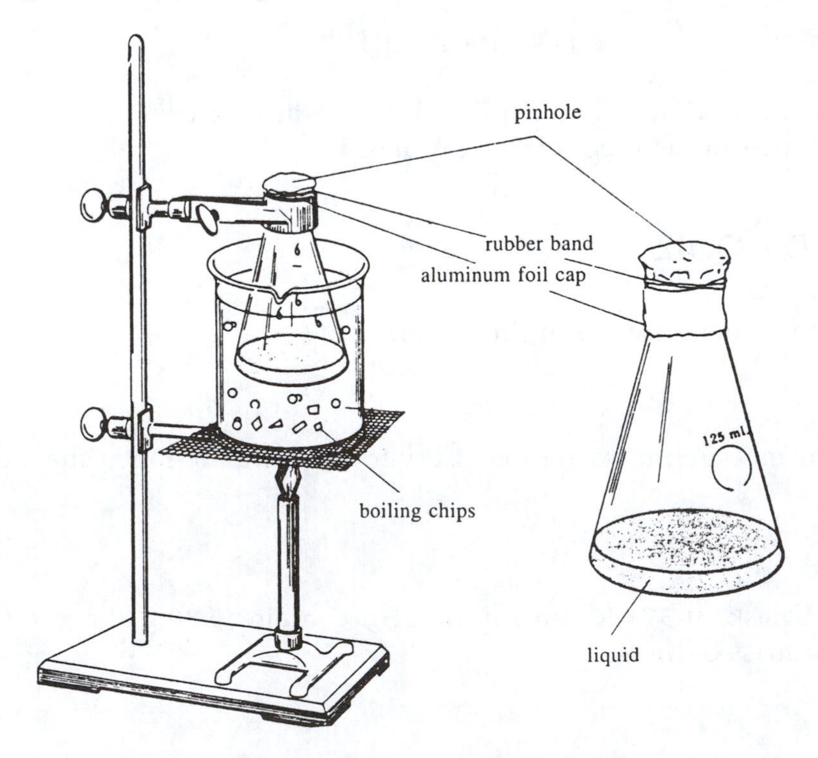

FIGURE 10-1

8. After the flask has cooled, you will observe some condensate inside the flask. Dry the flask on the outside, and again determine the mass of the flask and contents to the nearest 0.01 g (0.0001 g if you use an analytical balance). The difference between the first and second weighings of the flask is the mass of the vapor that filled the flask.

9. Repeat this determination with a second sample of the same unknown liquid.

10. After determining the mass of the flask and condensate the second time, take off the cap, empty out the liquid, fill the flask completely full of water, and determine the volume of the flask by pouring the water into a large graduated cylinder.

DATA RECORD

	Trial 1	*Trial 2*
Mass of flask, foil, and condensed vapor	________	________
Mass of flask (empty) and foil	________	________
Mass of vapor	________	________
Temperature of water	________	________
Barometric pressure	________	
Volume of flask	________	
Molar mass of vapor	________	________

Calculations (show all work)

Name ______________________ Class ____________ Room ____________

EXERCISE 11
A LOOK AT SOME INORGANIC CHEMISTRY

INTRODUCTION

Groups of elements with similar chemical properties are called families. Fluorine, chlorine, bromine, and iodine form such a family of elements called the halogens. Halogens and halogen compounds are widely used in industry and in medicine. Fluorine compounds are used to prevent tooth decay. They are also used to etch glass. Chlorine compounds are used in both bleach and in disinfectants. They are also found in table salt and in stomach acid. Bromine compounds are used in sedatives and in photographic emulsions. Iodine compounds are used in the treatment of goiter and as antiseptics. The halogens, except for fluorine, can be prepared by the oxidation of the halide salt. You will prepare them using manganese dioxide as the oxidizing agent.

Sulfur is used as a fungicide and in the manufacture of rubber. Sulfur exists in several different forms called allotropes. Sulfuric acid, H_2SO_4, is one of the most important industrial chemicals. It is used in the refining of petroleum and in the manufacture of fertilizers, explosives, dyes, and rayon.

The distinguishing feature of many metals is their ability to form complex structures called coordination compounds or metal complexes. Figure 11-1 shows two examples. Coordination compounds are widespread both in nature and in industry. Chlorophyll is a magnesium coordination compound. Hemoglobin, the compound responsible for carrying oxygen in the blood, is a coordination compound of iron. Many enzymes are metal complexes, as are most naturally occurring dyes.

In addition to preparing the metal complex shown in Figure 11-1a, you will study a metal complex that is involved in an equilibrium reaction. Most chemical reactions can be reversed under suitable conditions. When both the forward and the reverse reactions occur at the same speed, the reactions are said to be in equilibrium. The equilibrium you will study is shown in the equation that follows. Color changes in the solution make it easy to follow the shifts in the equilibrium.

$$\underset{\text{light yellow}}{Fe^{3+}} + \underset{\text{colorless}}{CNS^-} \rightleftharpoons \underset{\text{deep red}}{Fe(CNS)^{2+}}$$

(a) $[Cu(NH_3)_4]^{2+}$: Cu^{2+} bonded to four NH_3

(b) $[Fe(H_2O)_6]^{3+}$: Fe^{3+} bonded to six H_2O

FIGURE 11-1

PRELAB STUDY QUIZ

1. What are three uses for sulfuric acid?

2. Give two important examples of naturally occurring coordination compounds.

3. Find an example of an equilibrium reaction that occurs in your body.

4. Give two uses of fluorine compounds; of chlorine compounds; of bromine compounds; of iodine compounds.

PROCEDURE

Record all observations on the Data Record.

A. The Halogen Family

Caution: Do this experiment in the hood. Halogen vapors are toxic. Avoid breathing them.

Add a small amount of manganese dioxide (about 0.1 g) to each of three test tubes. Cover the bottom of the first tube with crystals of sodium chloride, the bottom of the second test tube with sodium bromide, and the bottom of the third with sodium iodide. Add a few drops of concentrated sulfuric acid to each and observe the reaction. Record all observations.

B. Sulfur

Place about 0.5 g of powdered sulfur in a test tube and slowly heat until the sulfur melts and just begins to boil. ***Do this in the hood.*** Record all changes that you observed. Add a few iron filings to the molten sulfur and observe the reaction. Allow the mixture to cool and then add a few drops of dilute hydrochloric acid to the contents of the test tube (***still under the hood***). Avoid breathing the vapors, which are very toxic. Hold a piece of wet lead acetate paper in the vapors coming from the test tube and record your observations. Appearance of a black compound on the strip indicates the formation of lead sulfide.

C. Preparation of a Copper Coordination Compound

Dissolve 10 g of powdered copper (II) sulfate (grind if necessary) in 25 mL of 12 M ammonium hydroxide. Stir until all the copper (II) sulfate has dissolved; then, while stirring, *slowly* add 15 mL of ethyl alcohol to the solution. Filter the deep blue crystals using a Buchner funnel and suction filter flask, as shown in Figure 11-2. Dry and weigh your crystals.

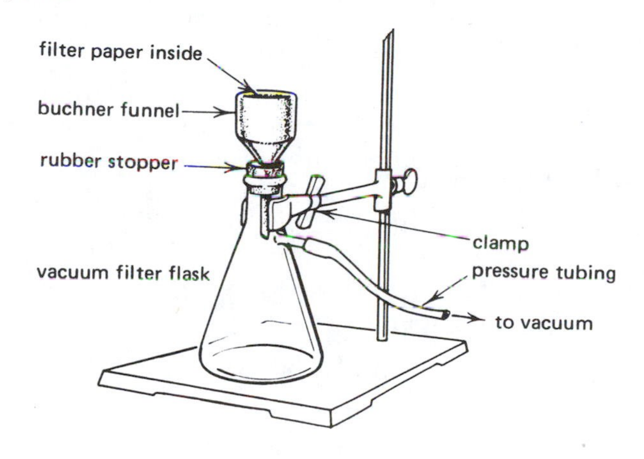

FIGURE 11-2

D. Equilibrium Involving an Iron Complex

Place 20 mL of water, 10 drops of 0.1 M iron (III) chloride, and 20 drops of 0.2 M ammonium thiocyanate into a small beaker. Divide the mixture equally among four test tubes. To the first tube add 5 drops of 5 M iron (III) chloride. To the second tube add 10 drops of 5 M ammonium thiocyanate. To the third add about 0.5 g of sodium fluoride. Compare the colors of these three test tubes with that of the fourth tube. Explain what happened on the Data Record.

DATA RECORD

A. Halogen Observations

Chlorine

Bromine

Iodine

B. Sulfur Observations

Heating sulfur

Addition of iron filings

Addition of hydrochloric acid

Use of moist lead acetate paper

C. Preparation of a Copper Coordination Compound

Describe the color changes that occur upon the addition of ammonium hydroxide.

What is the purpose of the ethyl alcohol?

Weight of copper coordination compound isolated. ______________________

Name ______________________ Class ____________ Room ____________

D. Equilibrium Involving an Iron Complex

Original color of mixture. ____________

Color upon addition of Iron (III) chloride. ____________

Color upon the addition of ammonium thiocyanate. ____________

Color upon the addition of sodium fluoride. ____________

Explain the cause for these color changes.

CASE HISTORY: HYPERKALEMIA

Ms. March was in an elevator accident and suffered crushing injuries of her legs. Emergency surgery was performed. She was given intravenously 100 mEq of potassium daily. After her surgery her urinary output diminished to almost nothing, but there was no bladder distention. On the second day after her surgery Ms. March appeared confused and complained of weakness and tingling sensations in her fingers and toes.

CLINICAL ASPECTS

1. Look at the following lab results.

MICHAEL REESE HOSPITAL AND MEDICAL CENTER

271830 (R-1-76)

UNIT RECORD NUMBER | V | B | MO | DAY | YR | N U | 1 | 2 | 3 | 4 | 5 | 6 | DEPT. CODE 603 | ☒ PRIVATE OUTPATIENT | ☐ EMERGENCY ROOM | ☐ MANDEL CLINIC

DIAGNOSIS — TOTAL CHARGE — JUNE MARCH

SPECIMEN — DATE AND TIME COLLECTED — DATE EXAMINED

RESIDENT OR INTERN: B. COHEN — REQUISITION PREPARED BY — TECH.

CHART

Test	Normal	Result	Test	Normal	Result	Test	Normal	Result	Test	Normal	Result
SODIUM ☒ 111	138-142 Meq/l	130	GLUCOSE ☐ 119 ☐ 2 Hr. Post Prandial	70-105 mg %		CALCIUM ☐ 124	9.5-10.5 mg %		AMYLASE ☐ 129	40-150 U.	
POTASSIUM 112 ☒	3.9-4.4 Meq/l	6.4	UREA N. 120 ☐	9-18 mg %		PHOSPHORUS 125 ☐	Ad.-2.5-4.0 Ch-4-7mg %		LIPASE 130 ☐	< 1.0 U.	
CHLORIDE ☒ 113	99-104 Meq/l	93	CREATININE ☐ 121	0.6-1.0 mg %		MAGNESIUM ☐ 126	1.8-2.3 mg %		Phenothiazines ☐ 191	Absent	
CO_2 CONTENT 114 ☒	26-29 mm/l	13	INSTRUCTIONS						Blood Ammonia 132 ☐	<150 µg %	
BLOOD pH ☐ 115	7.38-7.42								SALICYLATE ☐ 133	Absent	
BLOOD pCO_2 116 ☐	V-42-49mm A-38-41mm								BROMIDE 134 ☐	Absent	
BLOOD pO_2 ☐ 117	V-40 A->93		GLUCOSE TOL. ☐ 122	SEE SEPARATE REPORT		BILIRUBIN(T) ☐ 127	< 0.8 mg %		BARBITURATE ☐ 135	Absent	
ACETONE 118 ☐	<9mg %		CREATINE 123 ☐	0.5-1.2 mg %		BILIRUBIN(G) 128 ☐			DORIDEN 136 ☐	Absent	

(a) What laboratory tests were ordered?

(b) What are the normal ranges of each of the tests?

(c) Which results are outside the normal range?

Name ______________________ Class ____________ Room ____________

2. What are some common food sources of potassium?

3. What effects do corticosteroids have on potassium levels? Why?

4. Why is serum potassium measured rather than intracellular potassium?

5. How are potassium ions involved in the cardiac cycle?

6. How is potassium administered medicinally? What determines the amount to be administered? What determines the method of administration?

7. What are the clinical symptoms of hypokalemia? Hyperkalemia?

8. How does hypokalemia relate to the ECG?

9. How does hyperkalemia relate to the ECG?

BASIC SCIENCE ASPECTS

1. What causes hyperkalemia?

2. Hypokalemia may be caused by (a) too great a loss of potassium ions or (b) too little intake of potassium ions. Indicate possible causes of each.

3. What other effect(s) might cause hypokalemia?

4. Why do potassium ions have a great effect on cellular osmotic pressure? on cellular size?

5. What hormone(s) affect the concentration of potassium ions?

6. Compare active transport and diffusion of potassium ions. Why is each important?

7. What are the functions of potassium ions in the body?

8. If the serum potassium is 7.8 mg/100 mL, what is the number of milliequivalents per liter?

Name ______________________ Class ____________ Room ____________

Exercise 12
Water

INTRODUCTION

Except for oxygen, water is the most important substance known to man. It accounts for about two thirds of our body weight and plays an essential role in digestion, circulation, elimination, and regulation of body temperature. All cellular activities take place in a watery environment.

Water is a polar molecule and will dissolve only other polar materials. This helps explain why many ionic materials such as salts dissolve in water but nonpolar hydro-carbon materials such as gasoline and oil do not. The polarity of the water molecule also accounts for its surface tension. Surface tension has the effect of a thin, elastic membrane at the water's surface. When nonpolar materials such as dirt, oil, fats, or air come in contact with the surface of the water, it is difficult for them to penetrate this seemingly elastic membrane. Thus, detergents are needed to wash clothes and bile is needed for fat digestion. Both help mix the nonpolar materials with water.

Water is often found as an integral part of many salts; these salts are called hydrates. Most hydrates are stable when exposed to air at room temperature; however, some lose water when exposed to air. They are said to be efflorescent, whereas other salts will absorb moisture from the air and are said to be deliquescent.

You have already done a water purification experiment (Exercise 3), but now you will examine some of the dissolved materials found in water. Hard water contains calcium, magnesium, or iron ions. These ions interfere with the use of soap and leave a salt deposit behind when the water is boiled (boiler scale). To soften hard water, these ions must be removed. Water-softening units do this by replacing the offending ions with sodium ions. People on low-sodium diets, however, must be careful of water softened by this method.

PRELAB STUDY QUIZ

1. What function does bile perform?

2. What ions cause hardness in water?

3. Find three ways to soften hard water.

4. Why are small water bugs able to walk across a surface of water?

5. Diagram the structure of the water molecule, showing its polar nature.

6. Give the name and formula for two deliquescent hydrates, two efflorescent hydrates.

SPECIAL HAZARD

Avoid spilling silver nitrate on your skin. Although the skin discoloration (black) is not a health hazard, the results are considered unsightly.

PROCEDURE

Record all results on the Data Record.

A. Surface Tension

Into four separate clean test tubes placed about 5 mL of the following liquids: distilled water, 0.10% sodium chloride, 0.1% Ivory soap, and 0.1% bile salt. Set the tubes into the rack and carefully dust a small amount of sulfur onto the surface of each. Tap the sides of the tubes and record your observation. Which liquids allow sulfur to break through the surface?

B. Water of Hydration

Number four dry test tubes and place them in a test tube rack. Add 2 g of sodium chloride to the first tube; 2 g of copper(II) sulfate hydrate to the second, 2 g of aluminum sulfate hydrate to the third, and 2 g of barium chloride hydrate to the fourth. Heat the bottom of each tube in a Bunsen flame. Hold the test tubes horizontal while heating and watch for condensation in the cooler areas of the tube. Record the relative amounts of condensation.

C. Efflorescence and Deliquescence

Place several crystals of sodium sulfate on one watch glass and crystals of calcium chloride on another watch glass. Leave the crystals exposed to the air until next laboratory period. Describe the changes.

D. Dissolved Inorganic Content

1. *Chloride Content.* Add 3 drops of 1 M sliver nitrate and 1 drop of 1 M nitric acid to 5 mL of tap water in a test tube. Cloudiness indicates the presence of chloride ion. The amount can be qualitatively estimated as faint, light, medium, or heavy, depending upon the degree of cloudiness. Record your results.

2. *Phosphate content.* Add 1 drop of dilute nitric acid to 5 mL of tap water in a test tube. Check with litmus paper to make sure the solution is acidic. Add 5 drops of 0.1 M ammonium molybdate and boil the solution. A yellow precipitate indicates the presence of phosphate. Repeat this test using the following in place of tap water: (1) 1% solution of Calgon, (2) 1% solution of Tide or All, and (3) 1% solution of phosphate-free detergent. Record your results.

E. Determination of the Hardness of Water (Calcium Content)

Pour 20 mL of standardized calcium chloride solution (0.002 m) and 50 mL of distilled water into a 250-mL Erlenmeyer flask (see Figure 12-1 on page 72). Fill a buret with a stock soap solution and record the starting point. Add 0.3 mL of soap solution to the flask and shake. Repeat this procedure of 0.3 mL additions and shaking until a stable soap layer forms. Record the total volume used. Your instructor will tell you the concentration of calcium ions in the standardized calcium solution. Now repeat the preceding experiment using 20 mL of tap water instead of standardized calcium chloride solution. Record your results.

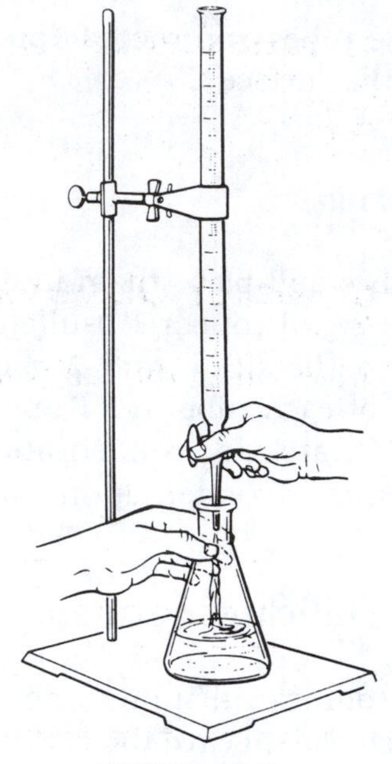

FIGURE 12-1

DATA RECORD

A. Surface Tension Observations

Distilled water

Sodium chloride

Ivory soap

Bile salt

Which liquids have low surface tension?

How do bile salts help food digestion? (See text.)

B. Water of Hydration Observations

Sodium chloride

Copper(II) sulfate hydrate

Aluminum sulfate hydrate

Barium chloride hydrate

Name ______________________ Class ____________ Room ____________

C. Efflorescence and Deliquescence Observations

Sodium sulfate

Calcium chloride

D. Dissolved Content of Water Observations

1. Relative amount of chloride ion in the tap water.

2. Phosphate content of

   Tap water ____________________

   Calgon ____________________

   Tide or All ____________________

   Phosphate-free detergent ____________________

E. Determination of Hardness in Water

1. Concentration of calcium in standardized calcium chloride solution ____________________

2. Grams of Ca^{2+} in 20 mL of solution ____________________

3. Volume of soap used in titration of standardized solution ____________________

4. Grams of Ca^{2+} needed to react with 1 mL of soap (answer 2 ÷ answer 3) ____________________

5. Volume of soap used to titrate tap water ____________________

6. Grams of Ca^{2+} in 20 mL of tap water (answer 4 × answer 5) ____________________

7. Grams of Ca^{2+} per mL of tap water (answer 6 ÷ 20 mL) ____________________

QUESTIONS

1. Why soften water?

2. Is rain water hard or soft? Why?

3. What is "deionized water"? How is it prepared? Where is it used?

4. In what units is hardness of water expressed?

5. Under what conditions should patients avoid drinking softened water?

Name ______________________ Class ____________ Room ____________

EXERCISE 13
SOLUTIONS

INTRODUCTION

Although there are many different types of solutions, only aqueous (water) solutions will be examined in this exercise. This is because water is the most important biologic solvent. Blood and urine are two examples of water solutions that occur in the body. Materials traveling to and from the cell must exist in a watery medium. Many medications are given as water solutions.

There are two parts to a solution: the solute, the material that has dissolved, and the solvent, the dissolving medium. There are a variety of ways to express solution concentration (the ration of solute to solvent). Dilute means a small amount of solute dissolved in a solvent, where as concentrated means a large amount of solute dissolved in a solvent. Saturated solutions contain all the solute they can hold at a given temperature. Unsaturated solutions contain a smaller amount of solute than they can normally hold under the given conditions. Sometimes it is possible to dissolve an excess amount of solute in a solvent. This situation produces a supersaturated solution. Such solutions are not stable and if a surface for crystal formation is provided, the excess solute will separate from the solution. All of the preceding terms, however, are relative and are not precise. In this exercise you will examine three exact methods of expressing solution concentration: percent by weight or volume, molarity, and milliequivalents per liter. Look up their definitions in your text and be sure that you are familiar with them before attempting Part G.

Chemical and pharmaceutical firms frequently supply solutions that must be diluted before use. The following equation can be used to solve dilution problems as long as the concentration units remain the same.

$$\text{volume}_{(\text{initial})} \times \text{concentration}_{(\text{initial})} = \text{volume}_{(\text{final})} \times \text{concentration}_{(\text{final})}$$

Suppose that you have 200 mL of a 5% glucose solution and you want to dilute it to 2%.

$$200 \text{ mL} \times 5\% = V_f \times 2\%$$

$$\frac{200 \text{ mL} \times 5\%}{2\%} = V_f$$

$$500 \text{ mL} = V_f$$

You would take the 200 mL of glucose and dilute it to a total volume of 500 mL.

In addition to studying solution concentrations, you also will look at several properties of solutions and at some factors involved in the solution process.

PRELAB STUDY QUIZ

1. How many milliequivalents of Ca^{2+} are in 11.0 g of $CaCl_2$?

2. If 20 g of sodium chloride are dissolved in 500 mL of solution, what is the percent concentration? the molarity?

3. A doctor needs 5 mL of 10% acetaminophen (Tylenol) solution; however, only a 25% solution is available. How much of the 25% solution will he need to make 5 mL of 10% acetaminophen solution?

4. What is a saturated solution? A supersaturated solution?

PROCEDURE

Record all results on the Data Record.

A. Effect of Surface Area and Stirring on Rate of Dissolution

Place several large crystals of copper(II) sulfate in a mortar and grind with a pestle. Evenly divide the pulverized material into two test tubes. Place a large unground copper(II) sulfate crystal in a third test tube. Add 5 mL of water to each tube and place all three tubes in a beaker that contains boiling water. Stir the contents of the first tube. Continue heating and stirring (tube 1) until all its contents have dissolved. Record the time it took for dissolution in each of the tubes.

Name ______________________ Class ______________ Room ____________

B. Saturated Solutions

Put 2 mL of saturated sodium chloride solution into two different test tubes. Add a few crystals of sodium chloride to the first and a few crystals of sugar to the second. Is there a difference? Record your results.

C. Solubility and Saturated Solutions

Put 1 g of sodium carbonate into a small beaker and add 10 mL of water. Stir. If all the sodium carbonate does not dissolve, continue adding 10–mL portions of water until it does dissolve or the beaker is full. Repeat the experiment using 1 g of calcium carbonate. What can you say about the relative solubilities of these two carbonates?

D. Heats of Solution

Put 5 mL of distilled water into a test tube and record the temperature. Rapidly dissolve about 2 g of ammonium chloride in the water and measure the temperature again. Repeat the experiment using 2 g of sodium hydroxide. (***Caution.***) Do not touch with fingers. Record your results.

E. Supersaturated Solutions

Place 5 g of sodium thiosulfate in a test tube and add 15 drops of water. Place the test tube in a boiling water bath until the crystals are completely dissolved. Place a rubber stopper *loosely* in the mouth of the test tube and let the tube cool to room temperature. Ad one crystal of sodium thiosulfate to the tube and record your observations.

F. Diffusion, the Dissolving Process

Put about 100 mL of water into a 250–mL beaker. Carefully drop one crystal of potassium permanganate into the water. Do not stir. Occasionally observe the beaker over the next few minutes. Can you explain what is happening?

G. Standard Solutions

Determine how many grams of sodium chloride you need to prepare 250 mL of 5% sodium chloride solution. After you have calculated this amount, weigh it out and dissolve it in enough water to make 250 mL of solution. Now calculate how many milliliters of this 5% sodium chloride solution are needed to make 100 mL of a physiologic saline solution (0.9% sodium chloride). After you have calculated this amount, take the correct amount of 5% sodium chloride solution and dilute it to 100 mL.

How many grams of calcium nitrate are needed to make 100 mL of 0.5 M solution? Now weigh out the proper amount and add enough water to make 100 mL. How many milliequivalents of Ca^{2+} are in the 100 mL of 0.5 M calcium nitrate?

DATA RECORD

A. Effect of Surface Areas and Stirring on Rate of Dissolution

Length of time for crushed copper(II) sulfate to dissolve

with stirring ________________ ; without stirring ________________

Length of time for copper(II) sulfate crystal to dissolve ________________

B. Saturated Solutions

Did the salt dissolve? The sugar? Explain.

C. Solubility and Saturated Solutions

Amount of water needed to dissolve sodium carbonate. ________________

Amount of water needed to dissolve calcium carbonate. ________________

D. Heats of Solution

Temperature change upon addition of ammonium chloride. ________________

Temperature change upon addition of sodium hydroxide. ________________

E. Supersaturated Solutions

Observations after crystal of sodium thiosulfate is added.

F. Diffusion, the Dissolving Process

Observations with the potassium permanganate crystal.

Name ____________________ Class __________ Room __________

G. Standard Solutions

Weight of sodium chloride needed for 250 mL of 5% solution ____________

Milliliters of 5% sodium chloride solution needed for 100 mL of 0.9% solution ____________

Weight of calcium nitrate needed for 100 mL of 0.5 M calcium nitrate ____________

Number of mEq of Ca^{2+} in 100 mL of 0.5 M calcium nitrate ____________

Calculations

Name ______________________ Class ____________ Room ____________

EXERCISE 14
COLLIGATIVE PROPERTIES

INTRODUCTION

Colligative properties of solutions are those properties that depend only on the number of solute particles in the solution. Freezing point depression, boiling point elevation, and osmotic pressure are three commonly observed colligative properties. The concentration of a solution is often expressed as molality (m) because this concentration is independent of temperature. Molality is defined as

$$\text{molality} = \frac{\text{moles of solute}}{\text{1000 g of solvent}}$$

In this experiment the freezing point of cyclohexane will be determined. Two solutions will be prepared and the freezing points of these will be determined. Given the value for the freezing point constant for cyclohexane is K_{fp} = –20 °C/molal, then the molar masses of the compounds in the solutions can be calculated form the equation

$$\text{molality} = \frac{\text{change in the freezing point temperature}}{K_{fp}\ (-20\ ^{\circ}\text{C/molal})}$$

PRELAB STUDY QUIZ

1. List three colligative properties of solutions.

2. Why is molality used as the concentration term when determining colligative properties?

3. Calculate the freezing point of a 5 molal sugar in water solution. K_{fp} for water is −1.86 °C/molal.

SPECIAL HAZARD

Hydrocarbons are highly flammable and their vapors may be explosive in the air. Be very careful with flames.

PROCEDURE

1. Place a large, dry test tube in a 250–mL Erlenmeyer flask. Using a beam balance, weigh the flask and test tube and record the mass to the nearest 0.01 g. Use a graduated cylinder to add about 20 mL of cyclohexane to the test tube and determine the total mass of the flask, test tube, and contents.

2. Place the test tube with the cyclohexane and a thermometer into the 600-mL beaker containing an ice-water mixture. Stir until the cyclohexane freezes. The cyclohexane will freeze at about 6 °C.

3. Remove the test tube from the ice-water and clamp it to the ring stand. Stir constantly, and as melting occurs, record the temperature every 30 seconds until it has reached about 8° C. Estimate the value between the marks on the thermometer as carefully as possible.

4. Weigh and record to the nearest 0.01 g, the mass of approximately 0.4 g of naphthalene. Dissolve it in the cyclohexane in the test tube.

5. Add thawing salt to the ice-water bath to lower its temperature below 0 °C. Place the test tube with the solution in the ice-water bath and stir until the solution is frozen.

6. Remove the test tube from the ice bath and record the initial temperature. Stir the solution continuously and record the temperature every 30 seconds until it is about 6 °C.

7. Obtain an unknown from the instructor and repeat steps 4 through 6 using the unknown and a fresh sample of pure cyclohexane.

8. Discard all of the cyclohexane and cyclohexane solutions in the special container labeled organic liquid wastes.

Name ______________________ Class ____________ Room __________

Sample Calculation

Suppose 1.0 g of an unknown solid dissolved in 40 g of cyclohexane lowers the freezing point of cyclohexane to 10.0 °C. To calculate the molar mass of the unknown we must first calculate the molality of the solution.

$$\text{molality of the solution} = \frac{-10\ ^\circ\text{C}}{-20\ ^\circ\text{C/molal}} = 0.50\ \text{molal}$$

To calculate the moles of unknown

$$\frac{0.50\ \text{mole of unknown}}{1000\ \text{g of cyclohexane}} = \frac{\text{X moles of unknown}}{40\ \text{g of cyclohexane}}$$

$$X = 0.02\ \text{mole of unknown}$$

To calculate the molar mass

$$\text{molar mass} = \frac{1.0\ \text{g of unknown}}{0.20\ \text{mole of unknown}} = 50\ \text{g/mole}$$

DATA RECORD

Mass of flask, test tube, and cyclohexane ______________

Mass of flask and test tube ______________

Mass of cyclohexane ______________

Mass of paper and naphthalene ______________

Mass of paper ______________

Mass of naphthalene ______________

Data for Unknown

Mass of flask, test tube, and cyclohexane ______________

Mass of flask and test tube ________________

Mass of cyclohexane ________________

Mass of paper and naphthalene ________________

Mass of paper ________________

Mass of unknown ________________

MELTING POINT DATA

Pure Cyclohexane				Cyclohexane + Naphthalene			
Temp.	*Time*	*Temp.*	*Time*	*Temp.*	*Time*	*Temp.*	*Time*

MELTING POINT DATA

Cyclohexane + Unknown					
Temp.	*Time*	*Temp.*	*Time*	*Temp.*	*Time*

Plot the temperature vs. time for the pure cyclohexane and for each solution using the graph paper on page 86. Determine the freezing points for each sample.

Freezing point: pure cyclohexane ________________

Name ______________________ Class ____________ Room ____________

Freezing point: cyclohexane and naphthalene solution ____________

Freezing point: cyclohexane and unknown solution ____________

The K_{fp} for cyclohexane is –20.0 °C/molal

Calculations

Show your calculations for the cyclohexane and naphthalene solution.

Molar mass of naphthalene ____________

Show your calculations for the cyclohexane and unknown solution:

Molar mass of unknown ____________

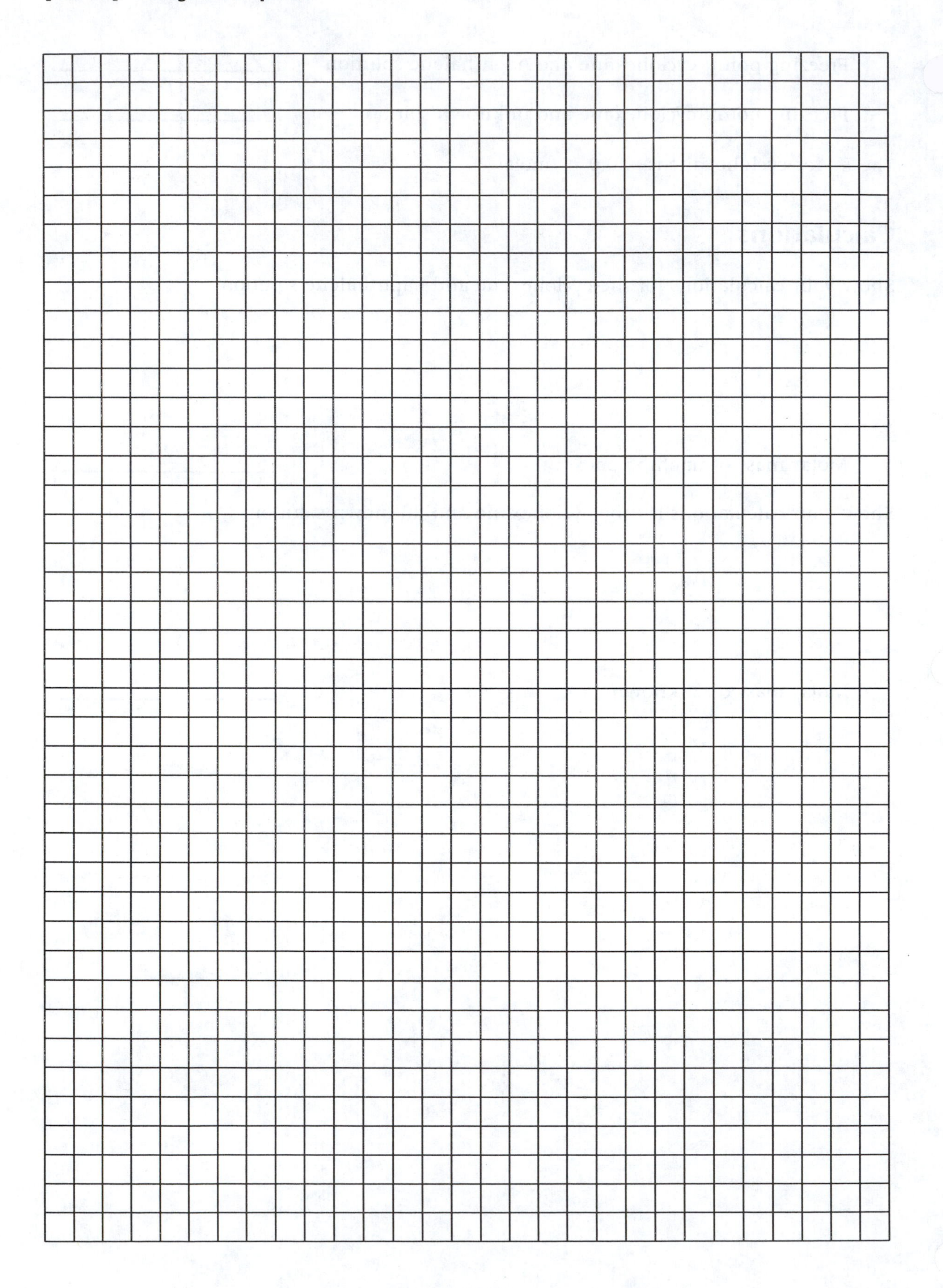

Name ______________________ Class ____________ Room ____________

EXERCISE 15
COLLOIDAL DISPERSIONS, OSMOSIS, AND DIALYSIS

INTRODUCTION

Colloidal dispersions involve particles whose size is larger than those found in solutions but smaller than those found in suspensions. Even though colloidal particles are insoluble, they remain suspended in the liquid. They do not settle, as do particles in a suspension. An example of a colloidal dispersion is the proteins carried in the blood stream. In this experiment you will prepare some colloidal dispersions and examine some of the properties of colloids. When a strong beam of light shines through a liquid, the Tyndall effect is exhibited as the larger-size colloidal particles reflect and scatter the light; smaller-size particles in solution do not.

The movement of water, ions, and small molecules throughout the body is of great importance. Osmosis and dialysis are two ways in which this type of movement is accomplished. In osmosis, water flows across a semipermeable membrane in order to equalize the solution concentration on both sides of the membrane. Water always moves from an area of lesser concentration to an area of greater concentration. The force or pressure across the membrane is called the osmotic pressure.

Dialysis is the separation of a solution from a colloidal (protein) by means of a semipermeable membrane. Such a membrane is called a dialyzing membrane. This type of membrane is used in artificial kidney machines.

PRELAB STUDY QUIZ

1. Find three common examples of colloidal dispersions.

2. Selective passage of water across a membrane is called ____________________.

3. Water always flows across a membrane from an area of ____________________ concentration to an area of ____________________ concentration.

4. Give examples showing the importance of dialysis and osmosis in the body.

SPECIAL HAZARD

Avoid spilling silver nitrate on your skin. Although the skin discoloration (black) is not a health hazard, the results are considèred unsightly.

PROCEDURE

Record all results on the Data Record

A. Osmosis and Osmotic Pressure

Carefully insert a thistle tube through a cork stopper. Tie a piece of cellophane over the large end of the tube and fill it two-thirds full with 20% sugar solution (see Figure 15-1.) Immerse the cellophane covered end of the tube into a beaker of distilled water until the water level and the sugar solutions levels are equal. Place a rubber band around the tubing to mark the initial level of the solution. Check every 15 minutes until the end of the period. Record the time and the distance (in millimeters) that the liquid level rises.

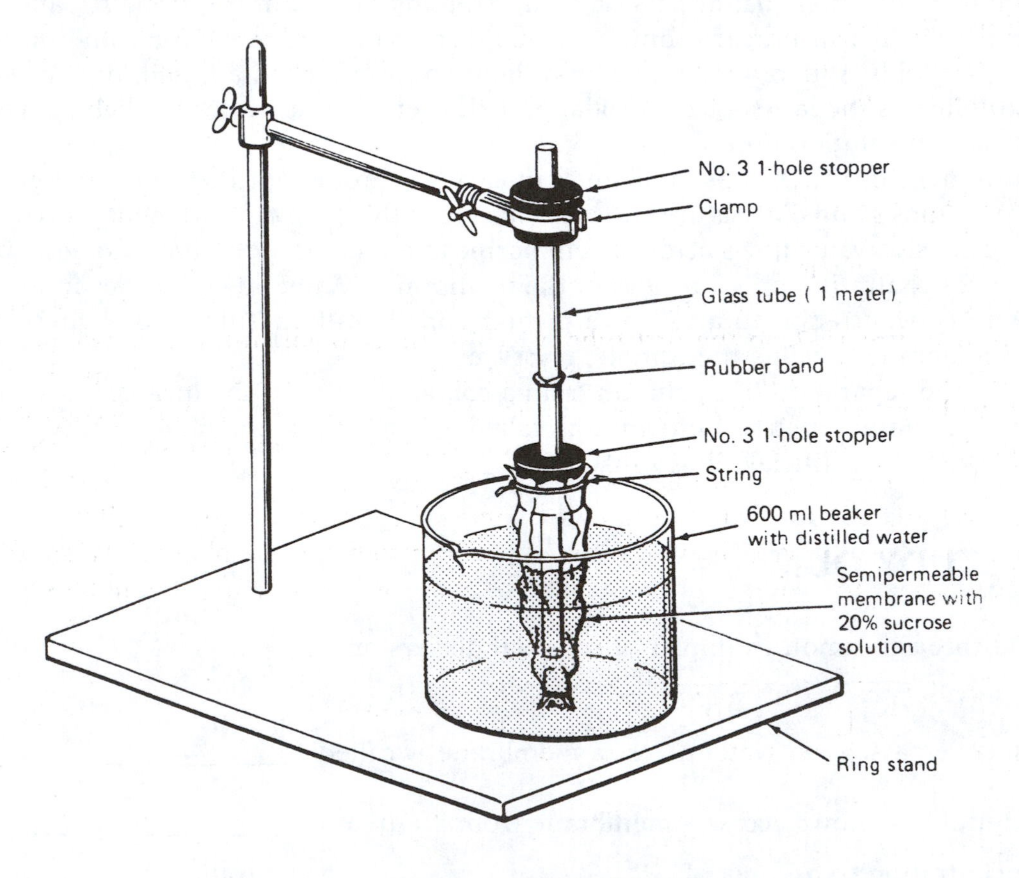

FIGURE 15-1

Name ______________________ Class ______________ Room ______________

B. Preparation of a Colloid

Boil abut 100 mL of distilled water. While the water is boiling, add a few drops of 10% ferric chloride solution. Colloidal ferric hydroxide forms.

C. Colloidal Dispersions and the Tyndall Effect

Prepare 0.1% and 0.01% starch dispersions as follows: Take 10 mL of the 1% starch dispersion provided and mix it with water until you have 100 mL. Then take 10 mL of that liquid and dilute it to 100 mL. Take five test tubes and fill each two-thirds full with one of the following liquids: 1% starch; 0.1% starch; 0.01% starch; 1% sugar; distilled water. Shine a light through each test tube. Observe the amount of light that travels through and the amount reflected at right angles. Which liquids did not give the Tyndall effect?

D. Preparation of a Gel

Place 15 mL of ethyl alcohol in a small beaker. As you swirl the alcohol, add all at once 2 mL of saturated calcium acetate. As soon as the calcium acetate is added, immediately cease swirling. Wait for a minute and then turn the beaker upside down. The material should be solid. Cut out a piece, put it on a wire gauze, and light it. Record your observations.

E. Dialysis

Cellophane tubing is a convenient dialyzing membrane. Take three 15-cm (6-in.) lengths of this tubing and puncture each near its end so that it can be supported as shown in Figure 15-2. In the first tube place 10 mL of potassium permanganate. In the second place 10 mL of prussian blue solution. In the third place 5 mL of 1% starch and 5 mL of sodium chloride solution. Be careful that none of the liquids contaminate the outside of the tubing. If they do, wash off with distilled water. Suspend each tube in a beaker of distilled water for 30 minutes. Test the liquid outside the starch–salt mixture for starch (iodine test) and for chloride ions (silver nitrate test). Record your results. Which substances passed through the membrane? What does this experiment tell you about the relative sizes of the particles in these liquids?

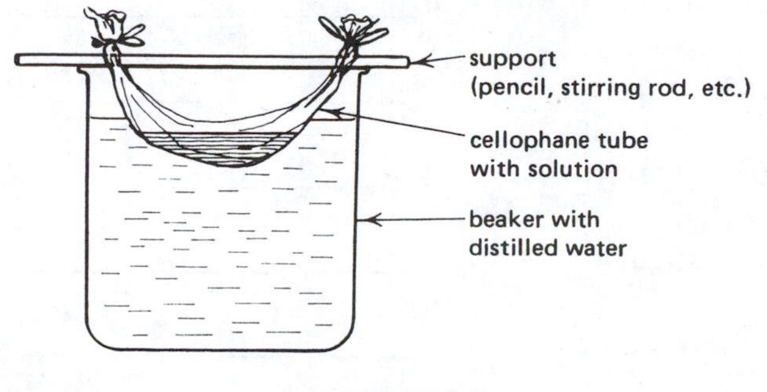

FIGURE 15-2

DATA RECORD

A. Osmosis and Osmotic Pressure

Time (minutes)	*Liquid Rise (mm)*

B. Preparation of a Colloid

Observations:

C. Colloidal Dispersions and the Tyndall Effect

Liquid	*Observations*
1% starch	
0.1% starch	
0.01% starch	
1% sugar	
distilled water	

Name ______________________ Class ____________ Room __________

D. Preparation of a Gel

Observations:

E. Dialysis

Did potassium permanganate diffuse through the membrane? Explain.

Did Prussian blue diffuse through the membrane? Explain.

Did the starch and the salt diffuse through the membrane? Explain.

Name ______________________________ Class ______________ Room __________

EXERCISE 16
ACIDS, BASES, AND SALTS

INTRODUCTION

Acids are compounds that furnish H^+ (actually H_3O^+) in solution. Acids have a sour taste. They react with many metals to form a salt and hydrogen gas; they react with carbonates and bicarbonates to form a salt, water, and carbon dioxide; they react with bases to form a salt and water. This last reaction is called neutralization.

Bases are compounds that react with acids by accepting hydrogen ions. Bases include not only hydroxides but also oxides, carbonates, bicarbonates, and ammonia. The word alkali is often used to mean such a base.

Salts are compounds in which the hydrogen of an acid has been replaced by a metal. Salts are formed by the reaction of an acid with a base.

Acids and bases may be classified according to their degree of ionization in solution. Strong acids and bases are highly ionized, whereas weak acids and bases are only slightly ionized. Common strong acids include nitric, sulfuric, hydrochloric, perchloric, hydrobromic, and hydroiodic. Most other acids are weak. Most hydroxide bases are strong, except for ammonium hydroxide, which is weak.

The degree of ionization of an acid or base may be determined by measuring the pH of the solution. Mathematically,

$$pH = -\log [H^+]$$

where $[H^+]$ is the concentration of hydrogen ions in the units moles per liter. Table 16-1 lists pH values and their corresponding $[H^+]$ and $[OH^-]$.

TABLE 16-1

pH	0	1	2	3	4	5	6	7	8	9	10	11	12	13	14
H^+	10^0	10^{-1}	10^{-2}	10^{-3}	10^{-4}	10^{-5}	10^{-6}	10^{-7}	10^{-8}	10^{-9}	10^{-10}	10^{-11}	10^{-12}	10^{-13}	10^{-14}
OH^-	10^{-14}	10^{-13}	10^{-12}	10^{-11}	10^{-10}	10^{-9}	10^{-8}	10^{-7}	10^{-6}	10^{-5}	10^{-4}	10^{-3}	10^{-2}	10^{-1}	10^0

At pH 7, the $[H^+]$ and the $[OH^-]$ are equal; the solution is neutral. Any solution with a pH of less than 7 is acidic; the lower the pH, the more acidic the solution. Solutions with a pH greater than 7 are basic; the higher the pH, the more basic the solution.

Large pH differences exist in the body. The gastric juice is very acidic, pH about 1. The pH of the intestinal contents varies from 2 to 6.6. Saliva is nearly neutral, its pH being near 7. Urine is slightly acidic, with a pH range of 4.5 to 7. Blood is slightly basic, with a pH of about 7.4. If the pH of the blood falls below 7.1 or rises above 7.8, death ensues.

pH is very important in many areas of the medical and biologic sciences. Bacteria grow only in culture media of certain pH; changes in pH may inhibit their growth or even destroy them. Tissue cultures must be maintained at pH values corresponding to those of body fluids or they will not survive. Most drugs are weak acids or bases and are present in the body in both the ionized and nonionized forms. The proportion of ionized to nonionized drug depends upon the pH of the solution in which the drug is dissolved. The nonionized molecules are more fat soluble and can diffuse more easily across cell membranes; ionized molecules are not fat soluble and are less able to diffuse across most body membranes.

Aspirin, for example, is a weak acid. It is absorbed into the blood stream, principally from the stomach and upper portion of the small intestine. In both of these areas the pH is low. At a low pH, aspirin does not ionize and, therefore, the amount of nonionized aspirin is high. To treat an aspirin overdose, the physician must convert the nonionized aspirin to its ionized form in order to slow down its absorption into the blood stream. Since aspirin is a weak acid, an alkali such as sodium bicarbonate is use to convert aspirin to its sodium salt, which is ionized.

Salt hydrolysis is the reaction between water and the ions in a salt that produces a change in pH. If a salt is derived from a weak acid and a strong base (sodium bicarbonate), hydrolysis will produce a basic solution. If a salt is derived from a weak base and a strong acid (ammonium chloride), hydrolysis will produce an acidic solution. If a salt is derived from an acid and a base of equal strength, the solution will be neutral (sodium chloride). Salts frequently are used to change the pH values of a variety of solutions: medications, culture media, cleansing agents, and swimming pools. Salts also play an important part in buffer solutions.

A solution that can maintain a relatively constant pH when limited amounts of acid or base are added to it is said to be buffered. Buffered solutions are found in all body fluids and maintain the proper pH for each fluid. Buffers consist of a weak acid and a salt of the weak acid. Carbonic acid and sodium bicarbonate form one of the main buffers in the bloodstream. This blood buffer functions as shown in the following equations.

$$NaHCO_3 + H^+ \rightarrow Na^+ + H_2CO_3$$

$$H_2CO_3 + OH^- \rightarrow HCO_3^- + H_2O$$

This buffer system is aided by two others in the bloodstream: the $HPO_4^{2-}/H_2PO_4^-$ buffers and the blood protein buffers. All these buffers function to maintain the pH of the blood near 7.4 by neutralizing such metabolic products as lactic acid and ammonia.

An easily demonstrated acid-base reaction that takes places in the stomach is the neutralization of excess stomach acid by antacid tablets. Excess stomach acid can cause indigestion and heartburn. When an advertisement claims that an antacid tablet consumes 47 times its own weight in excess stomach acid, it means that 1 g of that product will neutralize 47 g of 0.1 M hydrochloric acid. You can check this claim by performing an acid-base titration as described in this experiment.

Name ______________________________ Class ____________ Room ____________

PRELAB STUDY QUIZ

1. The reaction between an acid and sodium bicarbonate produces ______________, ______________ and ______________.

2. Give two examples of a strong acid; a weak acid; a strong base.

3. What is the $[H^+]$ and the $[OH^-]$ of a solution with a pH of 6?

4. Give an example of a salt whose hydrolysis will produce an acidic solution; a basic solution; a neutral solution.

5. Indicate one of the blood buffer systems and show how it neutralizes acid or base.

6. Explain the treatment for an overdose of aspirin.

PROCEDURE

Record all results on the Data Record.

A. pH and Hydrolysis (pH meters may be substituted for pH paper, if available.)

Determine the pH of 1 M hydrochloric acid and 1 M sodium hydroxide as follows: Place a small amount of each into separate test tubes. Rinse a stirring rod thoroughly with distilled water, dip it into one of the solutions, and then touch it to a piece of pH paper. Compare the color with those on the chart. Rinse the stirring rod again with distilled water and repeat for the other solution. By the same method determine the pH of the following: sodium acetate, sodium bicarbonate, sodium chloride, ammonium acetate, sodium carbonate, potassium bisulfate, sodium phosphate, potassium nitrate, ammonium chloride, distilled water, distilled water after you have blown through it for a couple of minutes with a straw. (This list may be modified by your instructor.)

B. Chemical Reactions of Acids

Into six test tubes place about 2 mL of 10% HCl. Into the first drop a piece of copper wire; into the second a piece of iron wire or a small nail; into the third a piece of zinc; into the fourth a piece of magnesium ribbon; into the fifth a marble chip; and into the sixth a small amount (tip of a spatula) of sodium bicarbonate. If no reaction occurs, warm gently. (*Do not boil.*) Test the gas evolved by holding a lighted splint at arm's length over the mouth of each tube. Did all the gases give the same test?

C. Buffers

Place 3 mL of each of the following liquids into separate test tubes: distilled water, 0.1 M $NaHCO_3$, 0.1 M NaH_2PO_4, carbonate buffer (0.1 M Na_2CO_3 and 0.1 M $NaHCO_3$), phosphate buffer (0.1 M NaH_2PO_4 and 0.1 M Na_2HPO_4). Using pH paper, determine the pH of each liquid. Then add 1 drop of 1 M HCl to each test tube and redetermine the pH. Which test tube showed the greatest change in pH? Which the least? Prepare five more test tubes as described, but this time add 1 drop 1 M NaOH instead of the HCl. Are the results the same as before?

D. Titration of Stomach Acid

Your instructor will demonstrate a titration first. Obtain a buret and rinse it three times, using 5 mL of 0.1 M NaOH each time. Fill the buret with the NaOH solution, making sure that all bubbles have been removed from the tip of the buret. Crush and weigh accurately (to the nearest 0.1 g) one antacid tablet. Carefully transfer all of the crushed tablet to a 250–mL Erlenmeyer flask and add exactly 100–mL of 0.1 M HCl (use a 100 mL volumetric flask if possible). After the tablet has dissolved as completely as possible, add 2 drops of bromphenol blue indicator and titrate with 0.1 M NaOH until the solution turns from yellow to just barely blue. Repeat with another tablet.

DATA RECORD

A. pH and Hydrolysis

Solution	*pH*
Hydrochloric acid, 1 M	________
Sodium hydroxide, 1 M	________
Sodium acetate	________
Sodium bicarbonate	________

Name ______________________ Class ____________ Room ____________

Solution	*pH*
Sodium chloride	________
Ammonium acetate	________
Sodium carbonate	________
Potassium bisulfate	________
Sodium phosphate	________
Potassium nitrate	________
Ammonium chloride	________
Distilled water	________
Distilled water after blowing through it	________
________	________
________	________
________	________

What kinds of salts make water acidic? basic? leave it neutral?

What effect does blowing through distilled water have on its pH? Why?

B. Chemical Reaction of Acids

Which materials produced a gas upon the addition of the HCl?

Which materials required heating before a gas was given off?

Which materials gave off a flammable gas?

What was the flammable gas?

Which materials gave off a nonflammable gas?

What was this gas?

Which materials gave off no gas at all?

C. Buffers

Solution	*pH*
Distilled water	______
Distilled water + HCl	______
Distilled water + NaOH	______
$NaHCO_3$	______
$NaHCO_3$ + HCl	______
$NaHCO_3$ + NaOH	______
NaH_2PO_4	______
NaH_2PO_4 + HCl	______
NaH_2PO_4 + NaOH	______
Carbonate buffer	______
Carbonate buffer + HCl	______
Carbonate buffer + NaOH	______
Phosphate buffer	______
Phosphate buffer + HCl	______
Phosphate buffer + NaOH	______

Name ______________________ Class ____________ Room ____________

Which solutions had the largest change in pH?

Which solutions had the smallest change in pH?

Write equations for the reaction of one buffer with an acid and with a base.

D. Titration

Name of Tablet ____________________

	Trial 1	*Trial 2*
Weight of Tablet	________	________
Final volume of NaOH	________	________
initial volume of NaOH	________	________
Volume of NaOH used	________	________
Volume of HCl consumed	________	________
Grams of HCl consumed	________	________
Grams of HCl consumed/gram of tablet	________	________

Calculations

The volume of HCl consumed by the tablet equals 100 mL minus volume of NaOH used. Assume 1 mL of 0.1 M HCl weighs 1 g, then grams of HCl consumed equals milliliters of HCl consumed. The weight of acid consumed by 1 g of antacid tablet will be given by

$$\frac{\text{weight of HCl consumed}}{\text{weight of tablet}}$$

Is your answer close to that claimed by the manufacturer?

Name ______________________ Class ____________ Room ____________

EXERCISE 17
PREPARATION OF SODIUM BICARBONATE AND SODIUM CARBONATE

INTRODUCTION

Sodium bicarbonate, $NaHCO_3$, and sodium carbonate, Na_2CO_3, are important industrial chemicals with many uses. Sodium bicarbonate is used as baking soda, as an antacid, in the manufacture of effervescent salts and beverages, and in gold plating. Sodium carbonate is used in the manufacture of glass, paper, soaps and detergents, and textiles, in the treatment of hard water, and in the removal of sulfur from industrial air pollution. Sodium bicarbonate and sodium carbonate are made industrially by the Solvay process.

A concentrated ammonia solution containing sodium chloride absorbs carbon dioxide

$$NaCl + H_2O + NH_3 + CO_2 \rightarrow NH_4^+ + HCO_3^- + Na^+ + Cl^-$$

Ammonia bicarbonate, NH_4HCO_3, is soluble in water, but sodium bicarbonate is not very soluble. Therefore as bicarbonate ions are produced, they precipitate as sodium bicarbonate instead of ammonia bicarbonate.

$$Na^+ + HCO_3^- \rightarrow NaHCO_3$$

The sodium bicarbonate can be separated by filtration. When the sodium bicarbonate is heated to about 270°C, it decomposes to sodium carbonate, water, and carbon dioxide.

$$2\,NaHCO_3 \xrightarrow[270\ ^\circ C]{} Na_2CO_3 + H_2O + CO_2$$

PRELAB STUDY QUESTIONS

1. Give three uses each for sodium bicarbonate and sodium carbonate.

2. What is the source of carbon in the manufacture of sodium carbonate and sodium bicarbonate?

3. Why do you isolate sodium bicarbonate instead of ammonia bicarbonate?

SPECIAL HAZARD

Ammonium hydroxide is caustic and very irritating to the eyes and the mucous membranes. Wash any affected area with large amounts of water and avoid breathing the fumes. Dry ice is very cold and may produce frost-bite if handled.

PROCEDURE (PARTS 1, 2, 3, 4 UNDER THE HOOD).

1. Under the hood, add 60 mL of concentrated ammonium hydroxide to a 125-mL Erlenmeyer flask. (Caution: Use concentrated ammonium hydroxide only under the hood. Avoid breathing the ammonia fumes.)

2. (Hood) To the ammonium hydroxide solution, add 14 g of fine sodium chloride crystals in small amounts. Swirl the solution after each addition to dissolve the sodium chloride.

3. (Hood) When no more sodium chloride will dissolve, carefully pour the saturated solution into a 250-mL beaker. Be careful NOT TO transfer any undissolved sodium chloride into the beaker.

4. (Hood) Add 40 to 50 g of crushed dry ice (solid carbon dioxide) in the form of several pieces to the solution. Use tongs or gloves to handle the dry ice. As the dry ice dissolves, sodium bicarbonate will precipitate as white solid. Wait until all the dry ice has disappeared.

5. Filter the precipitate using a suction filter apparatus (see Figure 11-2, page 63), collecting the sodium bicarbonate on the filter paper.

6. Dry the sodium bicarbonate by placing it on a watch glass over a beaker of gently boiling water. Continue to dry until the odor of ammonia can no longer be detected.

7. Take about 1 g of your sodium bicarbonate and place it in a test tube. Heat the tube gently in a burner flame. Progressively heat the tube hotter as the compound shrinks due to the loss of carbon dioxide and water. When no further reaction is seen, quit heating and allow the tube to cool.

8. Put some of the sodium carbonate into one marked test tube and some sodium bicarbonate into another marked test tube. Add 5 mL of water to both test tubes and shake gently to dissolve. Test both solutions with litmus paper and phenolphthalein solution. Record your observations.

9. Add a few drops of diluted hydrochloric acid to both solutions. Record your observations.

DATA RECORD

Describe the test of sodium carbonate

with litmus paper

with phenolphthalein

with HCl

Describe the test of sodium bicarbonate

with litmus paper

with phenolphthalein

with HCl

QUESTIONS

1. Why should you be careful not to transfer any solid sodium chloride in step 3?

2. What is the purpose of the dry ice?

3. Why is the sodium bicarbonate dried over boiling water?

CASE HISTORY: BURNS

A 27-year-old man was brought to the emergency room with second- and third-degree burns over 20% of his body after falling asleep while smoking in bed. An intravenous line was started with a Ringer's lactate infusion, and a Foley catheter was inserted to ensure urine output. A 10% mafenide acetate solution was applied to the burned area. Over the next several days the wounds were treated with debridement and mafenide acetate, and closed temporarily with homografts. Later autografts were used for final closure.

CLINICAL APPLICATIONS

1. Note the following laboratory reports on pages 104 and 105:

   (a) What tests did the doctor order?

   (b) What is the normal range for each of these tests?

   (c) Which results are outside the normal limits?

MICHAEL REESE HOSPITAL AND MEDICAL CENTER

UNIT RECORD NUMBER | V | B | MO | DAY | YR | N U | 1 | 2 | 3 | 4 | 5 | 6 | DEPT. CODE 603 | ☐ PRIVATE OUTPATIENT | ☐ EMERGENCY ROOM | ☐ MANDEL CLINIC

271831

DIAGNOSIS | TOTAL CHARGE

SPECIMEN | DATE AND TIME COLLECTED | DATE EXAMINED

RESIDENT OR INTERN | REQUISITION PREPARED BY | TECH.

CHART

Test	Normal	Result	Test	Normal	Result	Test	Normal	Result	Test	Normal	Result
SODIUM ☒ 111	138-142 Meq/l	132	GLUCOSE ☐ 119	70-105 mg % ☐ 2 Hr. Post Prandial		CALCIUM ☒ 124	9.5-10.5 mg %	9	AMYLASE ☐ 129	40-150 U.	
POTASSIUM 112 ☒	3.9-4.4 Meq/l	5	UREA N. 120 ☐	9-18 mg %		PHOSPHORUS 125 ☐	Ad.-2.5-4.0 Ch-4-7mg %		LIPASE 130 ☐	< 1.0 U.	
CHLORIDE ☒ 113	99-104 Meq/l	103	CREATININE ☐ 121	0.6-1.0 mg %		MAGNESIUM ☐ 126	1.8-2.3 mg %		Phenothiazines ☐ 191	Absent	
CO_2 CONTENT 114 ☐	26-29 mm/l		INSTRUCTIONS						Blood Ammonia 132 ☐	<150 μg %	
BLOOD pH ☐ 115	7.38-7.42								SALICYLATE ☐ 133	Absent	
BLOOD pCO_2 116 ☐	V-42-49mm A-38-41mm								BROMIDE 134 ☐	Absent	
BLOOD pO_2 ☐ 117	V-40 A->93		GLUCOSE TOL. ☐ 122	SEE SEPARATE REPORT		BILIRUBIN(T) ☐ 127	< 0.8 mg %		BARBITURATE ☐ 135	Absent	
ACETONE 118 ☐	<9mg %		CREATINE 123 ☐	0.5-1.2 mg %		BILIRUBIN(G) 128 ☐			DORIDEN 136 ☐	Absent	

(R-1-76)

Name ______________________ Class ______________ Room ______________

UNIT RECORD NUMBER | V | | B | MO | DAY | YR | H U | 1 | 2 | 3 | 4 | 5 | 6 | DEPT. CODE 606 ☐ MRHP ☐ PRIVATE OUTPATIENT ☐ EMERGENCY ROOM ☐ MANDEL CLINIC

MICHAEL REESE HOSPITAL AND MEDICAL CENTER 22817 (R.4-76)

DIAGNOSIS

DATE AND TIME COLLECTED | DATE EXAMINED

DR. | REQUISITION PREPARED BY | TECH.

CHART COPY

OUTPATIENT ONLY TOTAL CHARGE

☐ OR in AM

CBC ☐ 111
HB 112 ☐
HCT ☐ 113
WBC 115 ☐
DIFF. ☐ 116

122088

DATE	TEST & NORMALS	W.B.C. X 10^3 4.8-10.8	R.B.C. X 10^6 M 4.6-6.2 F 4.2-5.4	Hgb. gm. M 14-18 F 12-16	HCT. % M 42-52 F 37-47	MCV μ^3 84-99	MCH $\mu\mu g$ 27-31	MCHC % 32-36
				20	55			

Meta.	Myelo.	Promyelo	BLASTS	NRBC/100 WBC	POLY-CHROM.
POLY. (40-70%)	STAB. (0-6%)	LYMPH. (20-45%)	MONO. (2-8%)	EOS. (0-5%)	BASO. (0-1%)

SICKLE PREP ☐ 128
L. AL. PHOS. 131 ☐
G 6 PD ☐ 134
MALARIA 140 ☐
HEMOSIDERIN URINE ☐ 151
RETIC ☐ 117
PLATELETS 120 ☐
SER. WESTERGREN ☐ 122 GEN
SER. WINT-ROBE 122 ☐
LE PREP ☐ 124
COMMENTS

Reprinted by permission of Michael Reese Hospital and Medical Center, Chicago.

2. What is the difference between second- and third-degree burns?

3. What is debridement and why is it used?

4. What is the difference between homografts and autografts?

BASIC SCIENCE ASPECTS

1. What is Ringer's solution and why is it used?

2. What effects do burns have on electrolyte concentrations? Explain.

3. What kind of a compound is hemoglobin? Where is it produced in the body?

4. Describe the body's mechanism for the destruction of hemoglobin.

Name ______________________ Class __________ Room __________

Exercise 18
Oxidation-Reduction by Bleaching Solution

INTRODUCTION

Although oxidation can be defined as the gain of oxygen, a more general statement would define it as the loss of electrons. Substances that lose electrons during chemical reactions are said to be oxidized. Any substance that can bring about an oxidation is said to be an oxidizing agent.

Reduction is the opposite of oxidation and can be defined as the gain of electrons. Oxidation cannot take place without reduction. Any substance that can bring about the reduction of another substance is said to be a reducing agent. Therefore, the substance that is oxidized is the reducing agent and the substance reduced is the oxidizing agent.

Oxidation-reduction reactions are of great importance in chemistry and medicine. In the cell, all the reactions that result in energy production are oxidation-reduction reactions. For example, the cells oxidize glucose to carbon dioxide, water, and energy. Many antiseptics such as potassium permanganate, hydrogen peroxide, and iodine, are oxidizing agents. Stain removers act by oxidizing or reducing stains to form water-soluble substances. In this experiment, you will carry out an oxidation-reduction reaction with bleach. The oxidizing agent most commonly found in bleach is sodium hypochlorite, with the hypochlorite anion being the active constituent.

The strength of bleach is expressed as percent available chlorine and usually varies from 1% to 5% (that is, a 2% available chlorine bleach would contain 2 g of available chlorine per 100 mL of solution.) Available chlorine is determined by the amount of chlorine that can be produced from the hypochlorite anion by reaction with an acid.

$$ClO^- + 2H^+ + Cl^- \rightarrow \underset{\text{available chlorine}}{Cl_2} + H_2O$$

(ClO^- is reduced)
(Cl^- is oxidized)

You will determine the available chlorine by following two reactions:

(1) $\underset{\text{bleach}}{ClO^-} + 2\,\underset{\text{iodine ion}}{I^-} + 2\,H^+ \rightleftharpoons Cl^- + I_2 + H_2O$ (ClO^- is reduced) (I^- is oxidized)

(2) $I_2 + 2\,S_2O_3^{2-} \rightarrow S_4O_6^{2-} + 2\,I^-$ (I_2 is reduced) ($S_2O_3^{2-}$ is oxidized)

The amount of iodine produced is related directly to the amount of hypochlorite ion in the bleach.

PRELAB STUDY QUIZ

1. Find two specific examples of metabolic oxidation-reduction and explain why they are important.

2. The body oxidizes ethyl alcohol. What is the alcohol oxidized to?

3. What are the oxidation products of glucose in the cell?

SPECIAL HAZARD

Bleach can cause chemical burns and damage your clothes. It is also an eye irritant. Wash any affected area with water and keep your hands away from your eyes. Acetic acid can cause severe eye and skin burns. Wash the affected area with water.

PROCEDURE

Enter all results and calculations in the Data Record.

Pipet 2 mL of your bleaching solution into a 250–mL Erlenmeyer flask. (***Caution:*** *Do not pipet by mouth. Use a pipet bulb. Have your instructor demonstrate its use.*) Using a graduated cylinder, add 60 mL of 5% potassium iodide/acetic acid solution. Swirl to mix. Titrate by adding 0.05 M sodium thiosulfate from a buret until the bleaching solution in the flask turns to a pale amber color. At this point add 50 mL of distilled water and 2 mL of starch solution. Continue the titration very carefully until the dark blue starch-iodine color disappears. Record the amount of 0.05 M sodium thiosulfate added. Repeat the titration twice more. Calculate the amount of active ingredient, NaClO and compare it with the manufacturer's claim.

Name ______________________ Class ____________ Room ____________

DATA RECORD

	Trial 1	*Trial 2*	*Trial 3*
Final volume of $Na_2S_2O_3$	______	______	______
Initial volume of $Na_2S_2O_3$	______	______	______
Volume of $Na_2S_2O_3$	______	______	______
% NaClO	______	______	______

Calculations

% NaClO = volume of $Na_2S_2O_3$ used in titration X 0.200

QUESTIONS

1. In what part of the cell do most oxidation-reduction reactions occur?

2. Does oxygen enter into all metabolic reactions in the body? Explain.

3. Give the names and formulas for three disinfectants in common use in a hospital.

4. Why must oxidation always accompany reduction?

Name ______________________ Class ______________ Room ____________

EXERCISE 19

PREPARATION OF AN ALUMINUM COMPLEX: RECYCLING ALUMINUM

INTRODUCTION

As the amount of waste material multiplies, the awareness of finite resources and the need to recycle materials grows. One major contribution to the waste problems is the disposable aluminum beverage can. Not only are aluminum cans much more resistant to breakdown than steel cans, but aluminum manufacturing is very expensive. In this experiment you will convert an aluminum scrap can into a useful aluminum compound. The reactions are

$$\underset{\text{(from can)}}{2\ Al} + 2\ KOH + 6\ H_2O \rightarrow \underset{\text{in solution}}{2KAl(OH)_4} + \underset{\text{gas}}{3\ H_2}$$

$$16\ H_2O + 2\ KAl(OH)_4 + 4\ H_2SO_4 \rightarrow 2\ \underset{\text{alum}}{[KAl(SO_4)_2 \cdot 12\ H_2O]}$$

The final product $[KAl(SO_4)_2 \cdot 12\ H_2O]$ is called alum. Alum is a hydrated double salt of K^+, Al^{3+} and SO_4^{2-}. Alum is used in dyeing fabrics and making pickles.

PRELAB STUDY QUIZ

1. To what compound will your aluminum can be converted?

2. What are some uses for this compound?

3. What material is being used to dissolve the can?

SPECIAL HAZARD

Potassium hydroxide and sulfuric acid are very corrosive and can cause severe skin and eye damage. Wash the affected area with water and report any spills to the instructor.

PROCEDURE

Cut up an aluminum can into pieces. Weigh out about 1 g of aluminum and cut into very small pieces. Put the pieces into a 250 mL-beaker and add 50 mL of concentrated potassium hydroxide (*caution:* potassium hydroxide causes severe skin burns). Warm the solution in a *hood* until the aluminum dissolves. Tiny particles of paint and other impurities may remain undissolved. Since hydrogen is explosive, care must be exercised to ensure its rapid removal away from the flame. Keep the liquid level at 25 mL by adding water. After all the aluminum has dissolved, filter the solution while hot, through a small amount of glass wool.

Cool the solution to room temperature or below. With continuous stirring, *slowly* add 20 mL of 6 M sulfuric acid. This must be done slowly because the reaction will generate a large amount of heat. Large pieces of aluminum hydroxide will be present. Warm the solution until all the aluminum hydroxide dissolves. After the solution clears, cool the mixture in an ice bath for 20–30 minutes. Crystals of alum should form. These can be filtered and washed with an alcohol-water mixture. Dry and weigh your crystals. Dissolve a small amount of your product in 10 mL of water and test with pH paper.

DATA RECORD

Weight of aluminum scrap used ______________________

Weight of alum made ______________________

pH of alum solution ______________________

QUESTIONS

1. Why does the addition of sulfuric acid generate so much heat?

2. Why do alum solutions taste sour?

Organic Chemistry Exercises

Name ______________________ Class ____________ Room ________

EXERCISE 20
ORGANIC CHEMISTRY: THE CHEMISTRY OF CARBON COMPOUNDS

INTRODUCTION

Carbon compounds make up most compounds in the body. Proteins, carbohydrates, fats, vitamins, and hormones are all organic compounds. In order to understand how the body functions, it is necessary to know about these compounds and their reactions. This experiment is a very brief introduction to a few organic compounds.

The organic compounds called hydrocarbons can be broken into two major classes: (1) aliphatic hydrocarbons, in which the carbons are usually in chains; and (2) aromatics, in which the carbons have a cyclic or ring structure.

Organic compounds may also be subdivided into types, depending upon the presence of certain functional groups. Table 20-1 gives the functional groups that you will encounter in this experiment.

TABLE 20-1

Type of Organic Compounds Found in this Experiment

Name	Type Formula	Example	
Alcohol	R–OH	CH_3–OH	(methanol)
Ether	R–O–R	C_2H_5–O–C_2H_5	(diethyl ether)
Aldehyde	R–C(=O)–H	CH_3–C(=O)–H	(acetaldehyde)
Ketone	R–C(=O)–R	CH_3–C(=O)–CH_3	(acetone)
Acid	R–C(=O)–OH	CH_3–C(=O)–OH	(acetic acid)
Ester	R–C(=O)–OR′	CH_3–C(=O)–O–CH_3	(ethyl acetate)

PRELAB STUDY QUIZ

1. Before beginning this experiment look up in your text an application or use for each of the specific compounds that appear in Table 20-1.

2. What are the two major classes of hydrocarbons? What is the difference between them?

3. What is an alkane; an alkene; an alkyne? Give an example of each.

4. Draw the structures for: methane; benzene; phenanthrene.

SPECIAL HAZARD

Formaldehyde is very irritating to the mucous membrane and skin. Wash any affected area with a large amount of water and report any spills to the instructor. Avoid breathing if possible. Hydrochloric acid can cause severe skin burns. Wash any affected area with water and report spills to the instructor.

PROCEDURE

If structural model kits are available, you and your instructor may want to go through the following reactions using the models. Enter all observations and results on the Data Record.

A. Presence of Carbon

Cover the bottom of a test tube with sugar; then heat it, keeping the upper part of the tube cool. Record your observations.

Name ______________________ Class ____________ Room ____________

B. Alcohols

1. Pour about 2 mL of methyl alcohol into a test tube and test with pH paper. Is it an acid or base?
2. Add 5 mL of water to the test tube containing the methyl alcohol. Get a piece of copper wire about 22.5 cm (9 in.) long and spiral the bottom 2.5 to 5 cm (1 to 2 in.) around a pencil. Heat the copper spiral until it glows red. Now, quickly dip it into the mixture. Repeat this several times until a distinct odor appears.

C. Aldehydes and Ketones

Add 5 mL of Benedict's solution to two test tubes containing (1) 1 mL acetaldehyde; and (2) 1 mL of acetone. Warm gently in a water bath for a couple of minutes. Record the color changes.

D. Ethers

Put a small amount of phenol (about 0.1 g) into a test tube; then add 2 drops of concentrated sulfuric acid and 1 mL of methyl alcohol. ***Caution:*** Contact with phenol or concentrated sulfuric acid will cause sever skin burns and eye damage. Warm gently. Note the color. Record your observations.

E. Ethers

Add 5 mL of acetic acid and 2 mL of concentrated sulfuric acid into each of three test tubes. Cautiously note the odor of the following alcohols before adding them. To the first tube add 5 mL of isoamyl alcohol. To the second tube add 5 mL of octyl alcohol. To the third tube add 5 mL of benzyl alcohol. (Your instructor may substitute other alcohols.) Heat the three tubes in a boiling-water bath for 10 minutes. Note the odor after heating and describe the odor if possible.

F. Oxidation

Place 5 drops of 10% potassium permanganate in each of eight test tubes. To the first tube add 1 drop of concentrated hydrochloric acid and 3 mL of hexane. To the second tube add 1 drop of concentrated hydrochloric acid and 3 mL of hexene. To the third tube add 3 mL of ethyl alcohol. To the fourth tube add 3 mL of isopropyl alcohol. To the fifth tube add 3 mL of tertiary butyl alcohol. To the sixth tube add 3 mL of acetaldehyde. To the seventh tube add 3 mL of acetone. To the eighth tube add 3 mL of acetic acid. Mix and record your observations. A color change means an oxidation took place.

DATA RECORD

A. Sugar Observations

B. Observations of Methyl Alcohol

1. pH of methyl alcohol:
2. Observations for reactions of methyl alcohol with heated copper wire.

C. Observations of the Reaction of Aldehydes and Ketones with Benedict's Solution

1. Acetaldehyde

2. Acetone

D. Observations from the Preparation of Ether

E. Observations from the Properties of Acetate Esters

1. Isoamyl alcohol

2. Octyl alcohol

3. Benzyl alcohol

Name ______________________ Class ______________ Room ____________

F. Observation from Oxidation by Potassium Permanganate

1. Hexane

2. Hexene

3. Ethyl alcohol

4. Isopropyl alcohol

5. Tertiary butyl alcohol

6. Acetaldehyde

7, Acetone

8. Acetic acid

QUESTIONS

1. Write equations for the reactions in Part F.

2. Can all alcohols be oxidized?

3. Is methyl alcohol an acid or a base? Explain.

4. What is the function of the heated copper wire in the methyl alcohol reaction?

5. What is one use of esters?

Name ______________________ Class ______________ Room ____________

EXERCISE 21
DISTILLATION

INTRODUCTION

The space above a liquid will always contain some molecules of the material in the vapor (gas) state. A liquid left open to the atmosphere will, therefore, change into the vapor state, or evaporate. In a closed container, however, the liquid molecules will continue to escape only until the space above the liquid becomes saturated with vapor molecules of the material. This saturation point is measured in terms of vapor pressure, being different for different liquids. The stronger the attractive forces between molecules in a liquid, the lower the vapor pressure of that liquid. Conversely, the lower the attractive forces between the molecules in a liquid, the higher the vapor pressure.

Thus, the fact that ether evaporates much more rapidly than water is an indication that the attractive forces are much less between ether molecules than between water molecules. Vapor pressures of liquids vary directly with temperature so that the higher the temperature the greater the vapor pressure of a liquid. When the vapor pressure of a liquid reaches the atmospheric pressure, the liquid begins to boil. Distillation is the process of boiling a liquid and then condensing the vapor to the liquid state. Distillation is the principal method for separating and purifying organic liquids, providing that the liquids have different boiling points. Distillation is widely used in the manufacture of alcoholic beverages, gasoline and other petroleum products, and many chemicals.

The distillation process can be controlled by regulating the temperature of the vapors at the top of the distillation system. As the distillation proceeds, the temperature increases. This increase in temperature allows the observer to tell what compounds are being distilled. Sometimes, however, a constant-boiling mixture, or azeotropic mixture, will form. This is a mixture with identical liquid and vapor composition. It acts as a single, pure compound even though it is actually a mixture. When an azeotropic mixture is found, methods of purification other than simple distillation must be used. An example of an azeotropic mixture is one containing 95% ethyl alcohol and 5% water.

In this experiment you will distill a methyl alcohol-water solution.

PRELAB STUDY QUIZ

1. How are intermolecular attractive forces reflected in the boiling points of liquids?

2. How can a distillation be monitored so as to tell what substances are being distilled?

3. What is an azeotropic mixture? Give an example.

PROCEDURE

Set up a distillation apparatus as shown in Figure 21-1. Make sure that the flask and condenser are clamped and that all joints are tight (to prevent leaks). The bulb of the thermometer should be just below the opening in the side-arm. Attach the water lines to the condenser to that water enters at the bottom and exits at the top of the jacket.

Place about 100 mL of a 50% aqueous methyl alcohol solution in the distilling flask along with 2 or 3 boiling chips. Be sure that your instructor checks your apparatus before you begin.

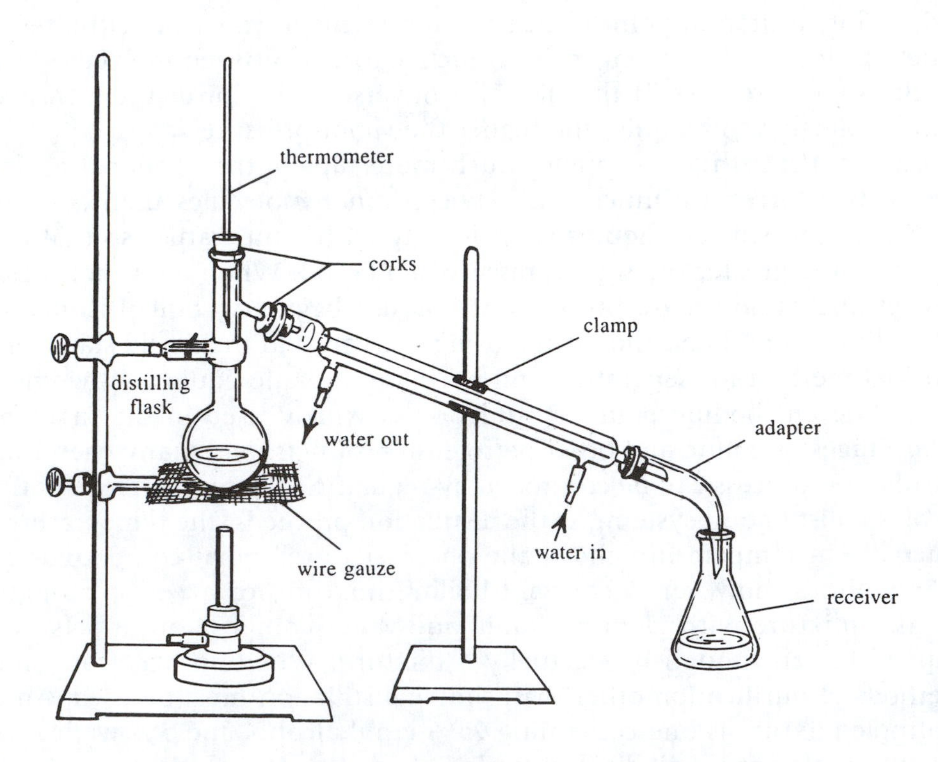

FIGURE 21-1 Diagram of distillation apparatus.

Name ______________________ Class ____________ Room ____________

Heat the distillation flask in such a manner that about one drop of distillate per second will be collected. Record the initial temperature of the first drop of distillate and then the temperature of each 5 mL interval of distillate, continuing until a total of 50 mL has been collected. Do *not* allow all of the liquid to be distilled.

DATA RECORD

Volume Collected	Temperature (°C)
First drop	
5 mL	
10 mL	
15 mL	
20 mL	
25 mL	
30 mL	
35 mL	
40 mL	
45 mL	
50 mL	

Observed boiling point for methyl alcohol ____________

Literature boiling point for methyl alcohol ____________

QUESTIONS

1. How would the observed temperature change if the thermometer bulb was located above the side arm?

2. If an unknown liquid is found to have a constant boiling temperature throughout an entire distillation, can you conclude that the liquid is a pure substance? Why or why not?

3. Two students in Denver, Colorado (elevator 5300 ft) distill the same pure liquid. One reports a boiling point of 10°C below that of the literature value for that liquid whereas the other reports a boiling point 2°C above the literature value. Which student reported the correct boiling point? Why?

Name ______________________ Class ____________ Room ____________

EXERCISE 22
EXTRACTION

INTRODUCTION

Extraction is one of several methods used to separate compounds in a mixture. Extraction involves the selective transfer of solute, or impurities, from one solvent to another. One of the solvents is usually water while the other is a water-insoluble organic liquid. Extraction is widely used to isolate compounds from natural sources. The process of brewing a pot of coffee or tea involves an extraction process.

The choice of solvents is based on the solubilities of the compound, or compounds, to be extracted. It is also based upon the volatility of the solvent since the extracted substance is usually isolated by distilling off that solvent. A single extraction usually does not remove one solute from a mixture. Therefore, several extractions are made. It has been found that several small extractions are much better than one or two large ones. Thus five 20-mL extractions are better than two 50-mL extractions, which in turn are better than one 100-mL extraction.

In this experiment you will use a sodium hydroxide solution to extract salicylic acid from a salicylic acid-naphthalene mixture dissolved in ether. The salicylic acid will be isolated by crystallization from an acidic solution while the naphthalene will be isolated by evaporating the volatile ether.

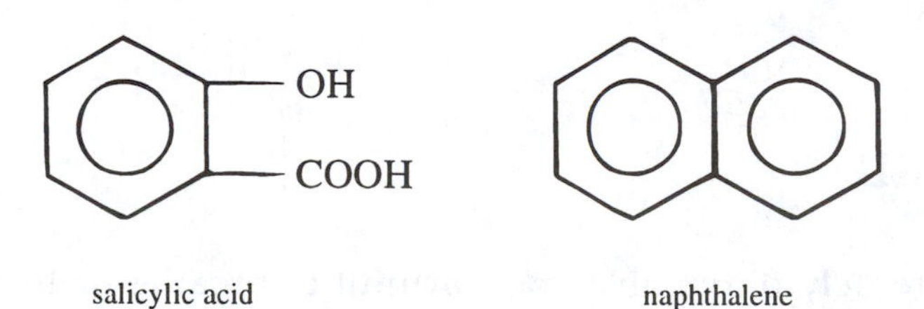

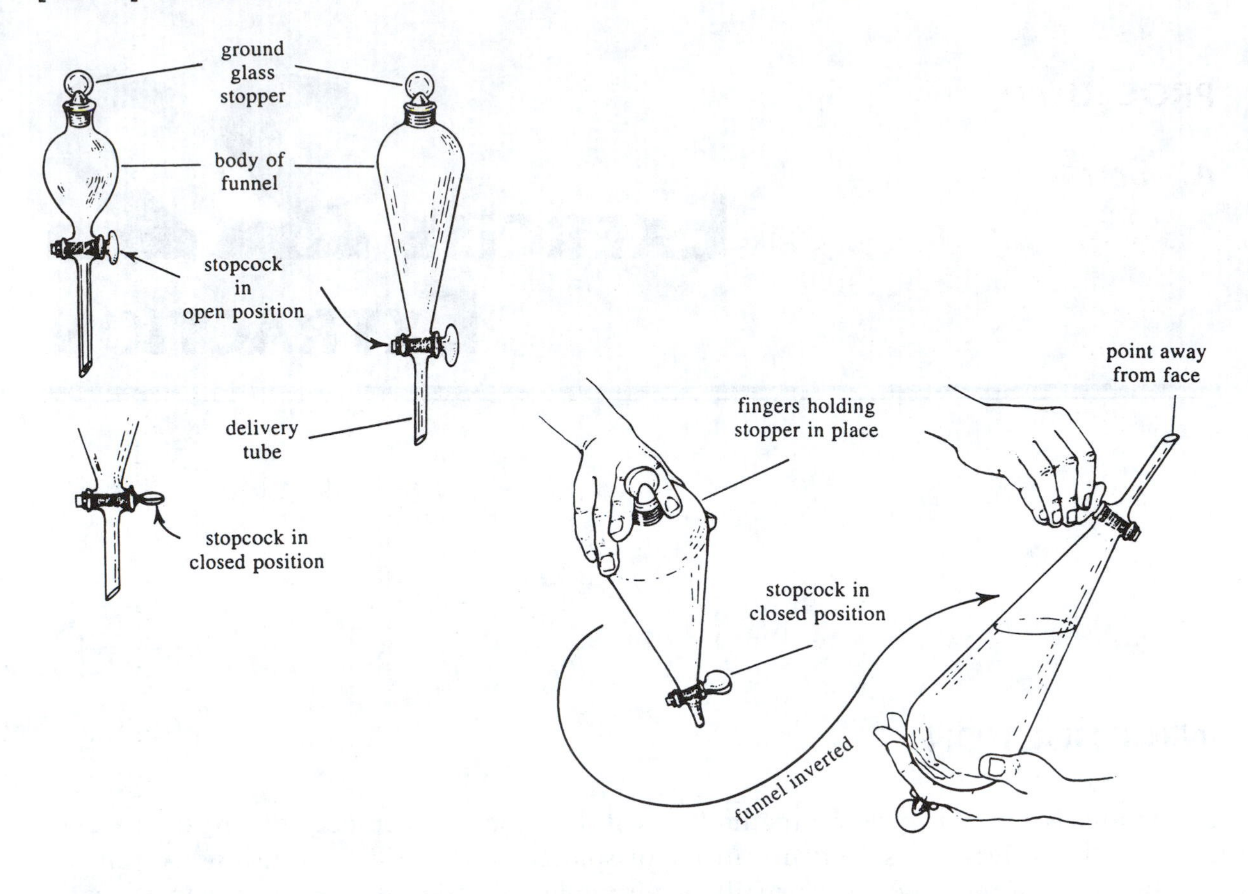

FIGURE 22-1

PRELAB STUDY QUIZ

1. What does volatility mean?

2. Give three applications for extraction.

3. Why in this experiment do we use two 25-mL sodium hydroxide extractions rather than one 50-mL extraction?

SPECIAL HAZARD

Diethyl ether is extremely flammable and harmful to breathe or touch. Keep ether far away from all flames or electrical devices.

PROCEDURE

A. Extraction

Weigh 5 g of a salicylic acid-naphthalene mixture in a 150-mL beaker. Add 50 mL of ethyl ether to the mixture. The mixture should dissolve in the ether. CAUTION: Be sure *all* flames are extinguished. Pour the solution into a 125-mL separatory funnel. Add 25-mL of dilute sodium hydroxide solution to the contents of the separatory funnel, stopper, invert, release the pressure, and briefly shake. Again invert and release the pressure. If you are unsure how to use a separatory funnel, have your instructor demonstrate its use. Continue shaking until no pressure buildup is perceived (approximately 4 or 5 shakings). Place the funnel upright in a ring on a ring stand. Remove the stopper and carefully drain off the bottom layer. Which layer is this? Which component of the mixture does it contain? Save the separated liquid. Repeat the extraction with another 25 mL of sodium hydroxide solution.

B. Separation

Combine both sodium hydroxide solution extracts (material removed from the separatory funnel) and warm gently on a hot water bath to drive off the ether. Cool the solution in an ice bath and add dilute hydrochloric acid until the solution becomes quite acidic to litmus paper. Acidification of the solution should produce a precipitate that can be isolated by suction filtration. The crystals thus separated should be washed with a small amount (5 to 10 mL) of cold water and then allowed to dry in the air (or in the lab desk). Once the crystals are completely dry, weigh them and save them for a melting point determination (see Part C). You may do this determination while waiting for the ether extract to dry.

Transfer the material remaining in the separatory funnel to a 125-mL Erlenmeyer flask, add 2 g of magnesium sulfate, and allow the mixture to stand for 20 minutes. Pour the liquid into an evaporating dish and place on a hot water bath in a hood. Allow the liquid to evaporate. Weight the material remaining, and determine its melting point.

C. Melting Point Determination

Place a small amount of the crystals in a melting point capillary tube. Attach the capillary tube to a thermometer with a small rubber band. Be sure that the thermometer can record temperatures up to 200°C. The bottom of the capillatory tube and the bottom of the thermometer should be together. Place the thermometer with attached capillary tube in an oil bath and heat slowly. Record the temperature at which melting occurs.

DATA RECORD

Weight of salicylic acid

Melting point of salicylic acid

Literature melting point

Weight of naphthalene

Melting point of naphthalene

Literature boiling point

QUESTIONS

1. What was in the bottom layer in the separatory funnel? Why was that layer at the bottom rather than the top?

2. Why is sodium hydroxide solution used to extract the salicylic acid rather than water?

3. What is the purpose of adding magnesium sulfate to the ether solution?

4. Write an equation for the reaction between sodium hydroxide and salicylic acid.

5. Why might the melting point you found be less than that in the literature?

Name ______________________ **Class** __________ **Room** __________

EXERCISE 23
PAPER CHROMATOGRAPHY

INTRODUCTION

One of the most common of all problems in chemistry is the separation of one substance from a mixture of others. While there are several methods that are used, few are better than chromatography. Chromatography is a separation technique defined as a method of separating components of a solution by distributing them between two immiscible phases. In paper chromatography the two phases are a stationary solid phase of paper and a moving liquid phase. If the components of a mixture differ in their attraction to the paper but have the same attraction to the liquid phase, then the components of the mixture will be selectively picked up by the moving phase as it passes over the stationary phase. Thus a separation is effected. In this experiment the separation of food coloring dyes will be effected by using a moving phase of isopropyl alcohol and water. The solid phase will be chromatography paper.

The ratio of the distance traveled by a compound to that traveled by the solvent s called the R_f factor.

PRELAB STUDY QUIZ

1. What is chromatography used for?

2. What are the phases in chromatography?

3. What is the moving phase in this experiment?

PROCEDURE

1. Prepare several micropipettes by heating and drawing out melting point capillaries to a diameter approximately that of a pin. The hole in the micropipettes should be small enough so that when it is touched to the surface of a solution a 1-cm-high column of liquid will be retained without forming drops.

2. Cut a 10-cm by 16-cm piece of chromatography paper and draw a pencil line about 1.5 cm from, and parallel to, the edge of the 10-cm side. Make evenly spaced pencil marks on this line and label as R (red), B (blue), Y (yellow), G (green) and M (mixture).

3. Using separate micropipettes for each color, make a small (1.5-mm) spot of the appropriate food color on the separate marks. Place all four on the M mark. Allow the spots to dry.

4. Roll the paper sufficiently to pass through the neck of a 500-mL Erlenmeyer flask. With the lower edge just touching the bottom of the flask, fasten the top of the paper to the inside of the flask with tape. The paper must be vertical and must not touch the sides of the flask.

5. Prepare a solution of 5 mL of water and 10 mL of isopropyl alcohol and carefully pour this solution into the bottom of the flask through a long-stem funnel. Do not splash. Cover the mouth of the flask with aluminum foil and allow the chromatogram to develop for about 30-45 minutes (until the solvent reaches the top).

6. Allow the chromatogram to dry and measure the R_f values.

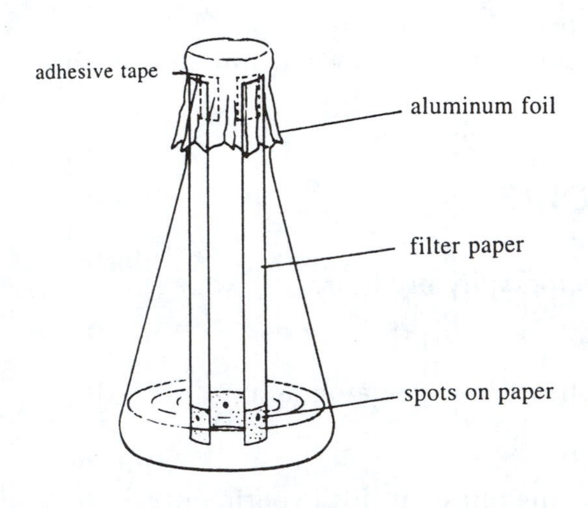

FIGURE 23-1

Name ______________________ Class ____________ Room ____________

DATA RECORD

	Distance traveled	R_f *value*
Red		
Blue		
Yellow		
Green		

Mixture

Red		
Blue		
Yellow		
Green		

Name ______________________ Class __________ Room __________

EXERCISE 24
POLYMERS

INTRODUCTION

Polymers are high molecular mass compounds in which many single molecules (monomers) are united to form giant molecules. Synthetic polymers often possess properties that are useful and mimic or improve upon the properties of such natural polymers as wool, cotton, cellulose (wool), and rubber.

Two principal types of reactions are used to prepare polymers: addition reactions in which double-bonded molecules are joined together, and condensation reactions in which functional groups are joined together with the accompanying production of byproducts such as water, ammonia, or hydrogen chloride. Polyethylene, polyvinyl chloride, "Teflon", synthetic rubbers, and polystyrene are all prepared by the first method. Nylon, polyesters, thiokol rubbers, proteins, and starch are examples of products of the second type of reaction.

You will prepare three polymers in this experiment: polystyrene, thiokol rubber, and nylon. Their equations for preparation are

Polystyrene

$$n\ \underset{\text{styrene}}{CH_2{=}CH(C_6H_5)} \xrightarrow[\text{peroxide}]{\text{benzoyl}} \sim CH_2{-}CH(C_6H_5){-}CH_2{-}CH(C_6H_5){-}CH_2{-}CH(C_6H_5) \sim$$

Thiokol rubber

$$NaOH + S \rightarrow n\ Na_2S_4$$

$$Na_2S_4 + n\ \underset{\text{ethylene dichloride}}{CH_2Cl{-}CH_2Cl} \rightarrow \sim CH_2{-}CH_2{-}S_4{-}CH_2{-}CH_2{-}S_4 \sim$$

Nylon

$$n\ \underset{\text{hexamethylene diamine}}{H_2N{-}(CH_2)_6{-}NH_2} + n\ \underset{\text{sebacoyl chloride}}{Cl{-}\overset{O}{\overset{\|}{C}}{-}(CH_2)_8{-}\overset{O}{\overset{\|}{C}}{-}Cl} \rightarrow n\ HCl +$$

$$\sim \overset{H}{\overset{|}{N}}{-}(CH_2)_6{-}\overset{H}{\overset{|}{N}}{-}\overset{O}{\overset{\|}{C}}{-}(CH_2)_8{-}\overset{O}{\overset{\|}{C}}{-}\overset{H}{\overset{|}{N}}{-}(CH_2)_6{-}\overset{H}{\overset{|}{N}}{-}\overset{O}{\overset{\|}{C}}{-}(CH_2)_8{-}\overset{O}{\overset{\|}{C}} \sim$$

PRELAB STUDY QUIZ

1. What does the reaction of sodium hydroxide with sulfur produce?

2. What are the two principal types of reactions used to prepare polymers?

3. What byproduct is produced when nylon is formed? When thiokol rubber is formed?

SPECIAL HAZARD

1,2-dichloroethane is flammable and harmful to breathe or touch. Keep 1,2-dichloroethane away from all flames. Benzoyl peroxide is a very strong oxidizer and can detonate if overheated on dry glassware. Follow instructions for its disposal and do not dump into waste cans. 1,6-hexadiamine and sebacoyl chloride are corrosive and sebacoyl chloride is a lachrymator. Report all spills to the instructor.

PROCEDURE

A. Preparation of Polystyrene

Place about 10 mL of styrene in a test tube and add 0.5 g of benzoyl peroxide. Mix to dissolve the solid and heat in a boiling water bath for an hour or more, or until the mixture has solidified or become quite viscous. If it is not yet solid, allow to stand and solidify at room temperature. Break the test tube and examine the contents.

B. Preparation of Thiokol Rubber

Heat 200 mL of 5% sodium hydroxide solution to boiling. Slowly add 15 g powdered sulfur. Stir until all the sulfur has dissolved and the solution has changed to a dark brown color. Allow the solution to cool to 70°C; add with stirring 0.5 to 1 mL liquid soap and 40 mL ethylene dichloride (1,2-dichloroethane). Continue to stir and maintain a temperature of 70°C until a yellow rubber product has formed. Remove, wash with water, and dry. Examine the product.

Name ______________________ Class ____________ Room ____________

C. Preparation of Nylon

Half fill a 100-mL beaker or vial with a 5% sebacoyl chloride solution in methylene chloride. Very slowly pour about 25 mL of a 5% aqueous solution of hexamethylenediamine over the methylene chloride solution. Do not mix or stir. Clamp the beaker or vial to a ring stand and notice the formation of nylon at the interface of the two layers. Using a copper wire with a curved tip, carefully begin pulling the nylon from between the layers. New nylon will be produced as the material is withdrawn from the container. Wind the nylon thread around some object. Do **not** touch it with your hands. When finished, wash the product thoroughly with water and test for strength by pulling on it. You may, when half done, thoroughly mix the contents in the beaker or vial by shaking. This will produce a lump of nylon. Remove, wash thoroughly, and test its strength.

QUESTIONS

1. What is the function of the benzoyl peroxide in the preparation of polystyrene?

2. Why should you avoid mixing the reagents when preparing nylon?

3. There is another isomer of ethylene dichloride. What is it? Show the formula of the thiokol rubber that would be formed if this isomer had been used.

Name ______________________ Class ______________ Room __________

EXERCISE 25
DRUGS

INTRODUCTION

What are drugs? Broadly defined, they are chemical agents that affect living processes. Many of us associate the term drug with substances that alter behavior (psychoactive drugs). Not all drugs are psychoactive. Caffeine and nicotine are considered to be psychoactive in low to moderate doses; aspirin is not.

Pharmacology, the study of drugs, embraces a wide area of study that is well beyond the scope of this manual. Medicine, however, is usually interested in the absorption, distribution, metabolism, excretion, and effects (both physiologic and psychologic) of a drug.

Let us take a look at one of the most common drugs in the United States, caffeine (Figure 25-1). Caffeine is found in coffee, tea, cola drinks, chocolate, and cocoa. Coffee and tea contain between 100 and 150 mg of caffeine per cup, whereas a 16-oz cola has between 45 and 70 mg of caffeine. Caffeine is also often an ingredient in many nonprescription drugs. A fatal dose of caffeine is unlikely (about 10 g), but serious side effects may be observed after ingesting a gram or more. (This dose corresponds to six to ten cups of coffee.)

Caffeine acts primarily by stimulating the central nervous system and the heart. It increases mental alertness, wakefulness, and restlessness (insomnia is frequently a side effect). It increases the total amount of blood pumped per minute by the heart and is used during congestive heart failure. Caffeine will also cause a constriction in the blood vessels leading to the brain and, for this reason, is used to treat migraine headaches. Caffeine acts as a mild diuretic. Recent evidence indicates that caffeine breaks down chromosomes and may have a deleterious effect during pregnancy.

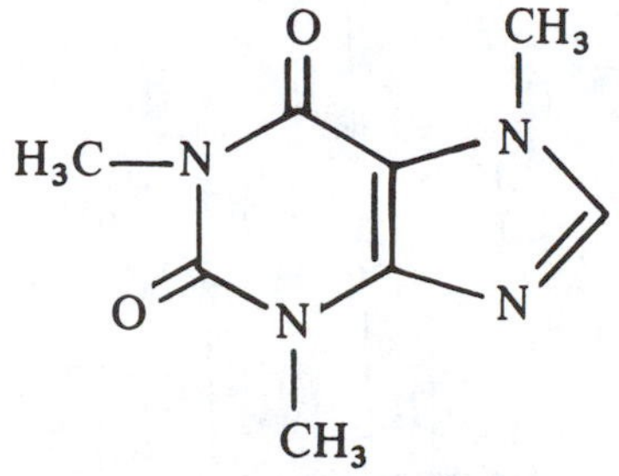

O CH_3
H_3C—N N
O N N
CH_3

caffeine

FIGURE 25-1

Aspirin and nicotine are two other widely used drugs. Aspirin (Figure 25-2) is used to reduce fevers, relieve headache pain, and reduce inflammation, especially from arthritis and rheumatic fever. Although the exact physiologic mechanisms of aspirin's action are not known, one important reaction may be aspirin's ability to bind metals. You will prepare a copper-aspirin in this exercise.

Nicotine (Figure 25-3) has no medical value, but its use is widespread since it is a constituent of tobacco. Nicotine is extremely toxic; cigarettes contain between 10 and 30 mg of nicotine, with about 10 percent (1 to 3 mg) actually absorbed into the blood stream).

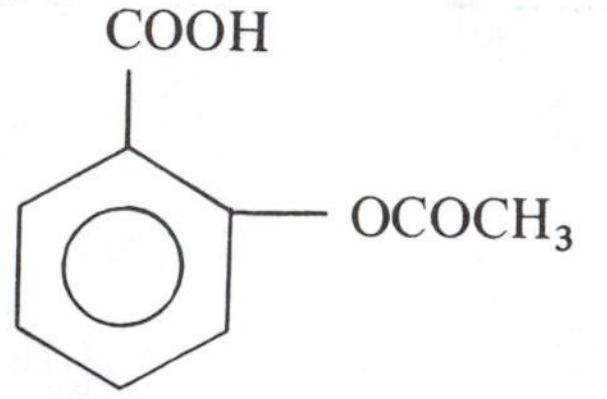

acetylsalicylic acid, aspirin

FIGURE 25-2

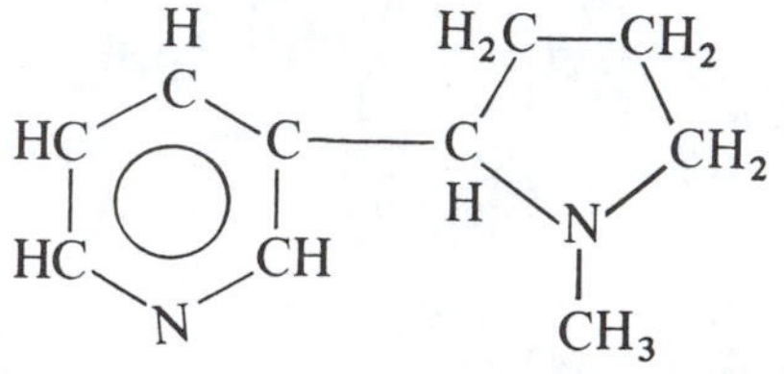

nicotine

FIGURE 25-3

Nicotine stimulates the central nervous system and increases heart rate and blood pressure. It is a bronchial irritant. In addition, nicotine, along with other constituents in cigarette smoke, is mutagenic and increases the rate of spontaneous abortion, stillbirths, and early postpartum death of infants.

You will identify nicotine by using thin-layer chromatography (TLC). In TLC there are a stationary phase, consisting of a thin layer of cellulose (or alumina, silica, etc.) on a glass or plastic slide, and a moving phase, consisting of a solvent and the components of a mixture (Figure 25-4). Each compound in the mixture is held back by the stationary phase to a different extent. Each component also has a different solubility in the solvent and moves forward at a different rate. This combination of holding-back factors and moving-forward factors results in a separation. Identification of the compounds is made by comparing known R_f values to those determined experimentally. R_f is defined as

$$R_f = \frac{\text{distance from the origin to the center of the spot}}{\text{distance from the origin to solvent front}}$$

R_f values for three compounds found in tobacco are nicotine, 0.80; nornicotine, 0.26; and anabasine, 0.48.

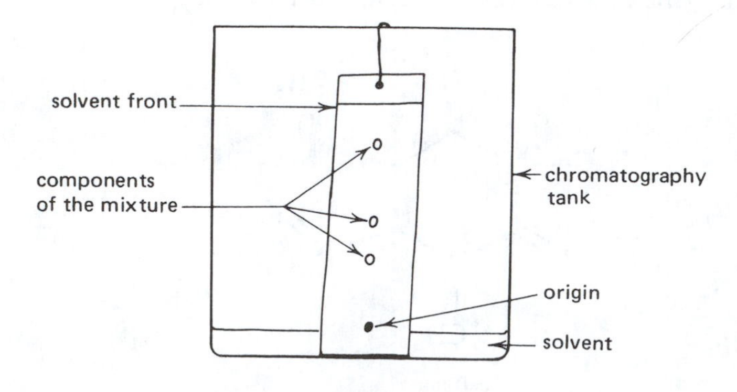

FIGURE 25-4

Name ______________________ Class ______________ Room ______________

PRELAB STUDY QUIZ

1. What are some of the most widely consumed drugs?
2. Give two medical uses for caffeine.
3. Give two medical uses for aspirin.
4. Give four medial problems caused by nicotine.

SPECIAL HAZARD

Acetic anhydride is very irritating to the eyes, nose, and throat. Avoid breathing its fumes. Methylene chloride is an anesthetic and a narcotic if inhaled in sufficient quantities. Avoid breathing or contact with methylene chloride. Toluene and petroleum ether are highly flammable. Keep them away from any flames. Nicotine is very toxic and any contact with the skin must be avoided. Report any skin contact to the instructor immediately.

PROCEDURE

Enter all observations and results on the Data Record.

A. Preparation of Aspirin

Place 2.5 g of salicylic acid in a 50-mL Erlenmeyer flask. Add 5 mL of acetic anhydride to the flask followed by 5 drops of 85% phosphoric acid. (***Caution:*** *Acetic anhydride can cause skin burns on contact. If you spill some on your hands, wash them immediately with water.*) Swirl the flask gently to mix the reagents. Place the flask in a 250-mL beaker containing about 150 mL of warm water (70 to 80°C) for 15 minutes. Without allow-

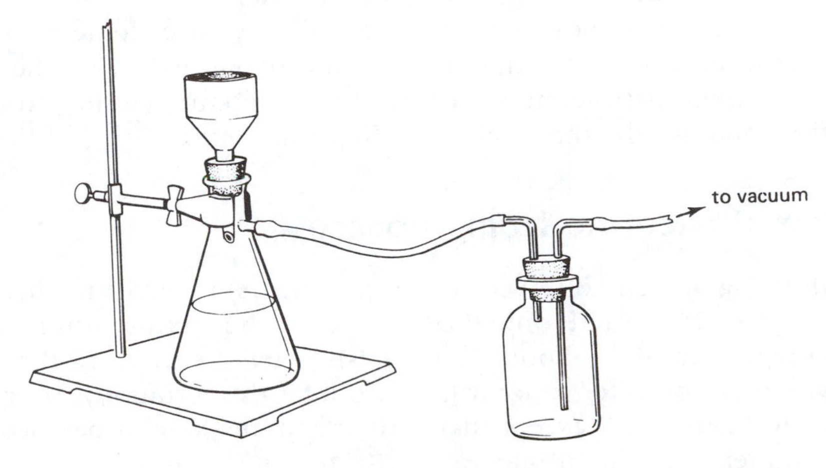

FIGURE 25-5

ing the contents to cool, carefully add, drop by drop, 2 mL of water to the flask. (***Caution:*** *Vigorous reaction accompanied by splattering may occur.*) When the reaction in the flask subsides, add 20 mL of water and let the flask cool to room temperature. At this point, place the flask in another 250-mL beaker that contains ice and water. Scratch the walls and the bottom of the flask with a stirring rod to induce crystallization. After crystallization is complete, filter the product through a Buchner funnel, as shown in Figure 25-5. Pour an additional 15 mL of cold water through the funnel to wash the product. Transfer the product to a dry piece of filter paper and allow it to dry at room temperature for at least 15 minutes. Weigh the finished product and record the weight.

B. Preparation of Copper Aspirinate

In a 100 mL beaker dissolve 1 g of aspirin in 20 mL of ethyl alcohol. In a 400-mL beaker dissolve 0.7 g of copper(II) acetate in 150 mL of cold water. Stir the solution until all the copper(II) acetate dissolves. Now pour the aspirin solution into the copper(II) acetate solution. Stir well and heat the mixture to 55°C. Set the mixture aside to cool and stir occasionally. Dark-blue crystals of copper(II) aspirinate will form upon cooling. Filter the crystals and wash with water. Your instructor will tell you what to do with them.

C. Isolation of Caffeine from Tea Leaves

Place 300 mL of water in a 500-mL Erlenmeyer flask, add 10 tea bags and 3.5 g sodium carbonate. Boil for 15 minutes. Cool the solution to room temperature and carefully add 20 mL of methylene chloride. The methylene chloride is used to extract caffeine from the tea. Transfer to a 250-mL separatory funnel and mix. (See page 126.) Your instructor will show you how to use a separatory funnel if you do not remember. Drain the methylene chloride/caffeine mixture (bottom layer) into a 100-mL beaker. Repeat the extraction process twice with two more 20-mL additions of methylene chloride and combine all the methylene chloride extracts.

Set up a steam bath, or warm water bath in the hood. Secure your beaker containing the methylene chloride and caffeine and allow all the methylene chloride to evaporate. (***Caution:*** *Avoid breathing toxic methylene chloride fumes.*) What is left in the beaker is a crude mixture. Weigh the crude caffeine.

If your instructor wishes, you may recrystallize your caffeine by dissolving it in 5 ml of warm toluene. Allow this solution to cool and follow by the drop-by-drop addition of enough petroleum ether to make the mixture cloudy. After it becomes cloudy, filter and dry the recrystallized caffeine in the air.

D. Isolation of Nicotine from Tobacco

Place about 10 g of tobacco (10 cigarettes, or 2 cigars) in a 250-mL beaker and pour about 100 mL of 25% NaOH onto the tobacco. Stir the mixture until all of the tobacco is thoroughly wetted (about 15 minutes). Filter the mixture through a Hirsch funnel with vacuum (filter paper will *not* work for this filtration). Your setup should be similar to Figure 25-5 except that a Hirsch funnel should be used instead of a Buchner funnel. Save the filtrate.

Name ______________________ Class ____________ Room ____________

Add 30 mL of methylene chloride to the filtrate and transfer to a 250-mL separatory funnel. Extract the nicotine into the methylene chloride layer. Ask your instructor to demonstrate if you do not remember. Drain the bottom layer of methylene chloride/nicotine into a 250-mL beaker. Repeat the extraction procedure with another 30 mL of methylene chloride, and combine both methylene chloride extracts.

At this point your instructor will tell you whether to do a qualitative test for nicotine by thin-layer chromatography.

E. Thin-layer Chromatography

Chromatography tanks containing a layer of chromatography solvent (methyl alcohol and methylene chloride) are provided. To prepare your sample for chromatography, obtain a melting point tube, open at both ends. Apply 1 drop of your nicotine solution to the thin-layer strip (about 2 cm from the bottom). Immediately dry the spot by blowing on it. Repeat this about 15 times, at same spot. Try to keep the spot as small as possible.

Put the strip into the tank and allow it to develop for 30 minutes or until the solvent gets near the top. Remove your strip, dry it, and expose the spot by using either ultraviolet light or a jar containing solid iodine. Mark the location of all spots located and determine their R_f values (see Introduction). Identify the composition of as many spots as possible by comparing R_f values obtained with those given in the Introduction or provided by the instructor.

DATA RECORD

A. Weight of aspirin obtained ____________

B. Weight of copper aspirinate obtained ____________

C. Weight of crude caffeine obtained ____________

D. Weight of recrystallized caffeine ____________

E.

Spot	*Distance from Origin (mm)*		R_f *Value*
Solvent	To Center of Spot	To Solvent Front	
1			
2			
3			
4			
5			

QUESTIONS

1. In the preparation of aspirin, why scratch the walls of the beaker?

2. How may the crude aspirin be purified?

3. Give a medical use for copper aspirinate.

4. What effect does caffeine have on chromosomes?

5. Explain how TLC works.

6. How does nicotine compare structurally with nicotinic acid? What is the latter used for?

Name ______________________ Class ______________ Room ______________

CASE HISTORY: ASPIRIN OVERDOSE

A 2½-year-old boy was brought to the hospital emergency room by his mother who had found him playing with an open bottle of aspirin. He was breathing rapidly and was perspiring profusely. On the way to the hospital he vomited once. The patient was given ipecac syrup to induce additional vomiting. Blood was drawn for testing of salicylate levels, and an intravenous infusion of 5% dextrose-½ normal saline was started. He was given sodium bicarbonate to alkalinize the urine and enhance excretion, and potassium chloride to replace losses. The boy recovered without complications.

CLINICAL APPLICATIONS

Note the following laboratory report.

1. What tests did the doctor order?

MICHAEL REESE HOSPITAL AND MEDICAL CENTER (R-1-76) 271817

UNIT RECORD NUMBER | V | B | MO | DAY | YR | N U | 1 | 2 | 3 | 4 | 5 | 6 | DEPT. CODE 603 | [ ] PRIVATE OUTPATIENT | [ ] EMERGENCY ROOM | [ ] MANDEL CLINIC

DIAGNOSIS | TOTAL CHARGE

SPECIMEN | DATE AND TIME COLLECTED | DATE EXAMINED

RESIDENT OR INTERN | REQUISITION PREPARED BY | TECH.

CHART

Test	Normal	Result	Test	Normal	Result	Test	Normal	Result	Test	Normal	Result
SODIUM [ ] 111	138-142 Meq/l		GLUCOSE [ ] 119 [ ] 2 Hr. Post Prandial	70-105 mg %		CALCIUM [ ] 124	9.5-10.5 mg %		AMYLASE [ ] 129	40-150 U.	
POTASSIUM 112 [ ]	3.9-4.4 Meq/l		UREA N. 120 [ ]	9-18 mg %		PHOSPHORUS 125 [ ]	Ad.-2.5-4.0 Ch-4-7mg %		LIPASE 130 [ ]	< 1.0 U.	
CHLORIDE [ ] 113	99-104 Meq/l		CREATININE [ ] 121	0.6-1.0 mg %		MAGNESIUM [ ] 126	1.8-2.3 mg %		Phenothiazines [ ] 191	Absent	
CO_2 CONTENT 114 [X]	26-29 mm/l	22	INSTRUCTIONS						Blood Ammonia 132 [ ]	<150 µg %	
BLOOD pH [X] 115	7.38-7.42	7,28							SALICYLATE [X] 133	Absent	70
BLOOD pCO_2 116 [X]	V=42-49mm A=38-41mm	A32							BROMIDE 134 [ ]	Absent	
BLOOD pO_2 [ ] 117	V=40 A=>93		GLUCOSE TOL. [ ] 122	SEE SEPARATE REPORT		BILIRUBIN(T) [ ] 127	< 0.8 mg %		BARBITURATE [ ] 135	Absent	
ACETONE 118 [ ]	<9mg %		CREATINE 123 [ ]	0.5-1.2 mg %		BILIRUBIN(G) 128 [ ]			DORIDEN 136 [ ]	Absent	

2. What are the normal ranges of these tests?

3. How did the lab results compare with the normals?

4. Why was KCl administered?

5. What is ipecac and why is it used?

6. What are the physiologic effects of aspirin?

7. When is the use of aspirin contraindicated?

8. What is the effect of an overdose of heroin? of vitamin A?

9. How may each of the conditions in Question 8 be treated?

BASIC SCIENCE ASPECTS

1. What is dextrose? How would you prepare a 5% solution of this substance?

2. Explain how $NaHCO_3$ can alkalize urine.

3. How does aspirin function in the body?

4. Chemically, what type of compound is aspirin? Is it soluble in water? in acid? in alkali?

5. What is a narcotic? an opiate? a psychedelic? Give two examples of each and indicate the effects of the use of each.

Name ______________________ Class ____________ Room ____________

EXERCISE 26
SYNTHESIS OF AN ARTIFICIAL FLAVOR

INTRODUCTION

Most flavors and odors in foods are due to organic compounds called esters. Since esters can be made in the laboratory, food, beverage, and fragrance companies have learned to use artificial esters to enhance their products. Natural odors and flavors are a result of complex mixtures of organic compounds. A careful blending of artificial compounds is necessary to successfully imitate a natural odor or flavor. For example, over twenty compounds have been identified in strawberry extract.

In this experiment you will prepare and purify an artificial odor/flavor.

Esters are made from the reaction of an alcohol with an acid as shown below.

$$\underset{\text{alcohol}}{R{-}OH} + \underset{\text{acid}}{H{-}O{-}\overset{\overset{\displaystyle O}{\|}}{C}R'} \overset{H^+}{\rightleftharpoons} \underset{\text{ester}}{R{-}O{-}C{-}\overset{\overset{\displaystyle O}{\|}}{C}{-}R'} + H_2O$$

This reaction seldom goes to completion and yields are usually low.

PRELAB STUDY QUIZ

1. What class of organic compounds are most flavors?

2. What two organic compounds are needed to make an ester?

SPECIAL HAZARD

Many of the organic acids in this experiment are foul smelling, toxic and irritating. Take special precautions to avoid spilling or breathing them. Immediately clean up any spills of organic acids. Sulfuric acid is very corrosive and will cause skin and eye damage. Wash any affected area with a large amount of water and report a spill to the instructor.

PROCEDURE

Choose a flavor to prepare from those listed in Table 26-1. Since several esters have high boiling points, check for the availability of high temperature thermometers first. The correct amount of acid and alcohol to use are given in the table. Butyric acid has a *very* disagreeable odor and should only be used in a well-ventilated hood.

After adding the correct amounts of acid and alcohol to a 250-mL round bottom flask, drop in 2 or 3 boiling chips and mix thoroughly. Slowly add 5 mL of concentrated sulfuric acid (*caution:* sulfuric acid causes severe burns). Assemble the apparatus shown in Figure 26-1.

TABLE 26-1

Flavor	Name of Ester	Amount of alcohol	Amount of acid	Boiling Point (°C)
Apple	methyl butyrate	10 g methyl alcohol	62 g butyric acid	102
Banana	isopentyl acetate	22 g isopentyl alcohol	36 g acetic acid	142
Orange	octyl acetate	33 g octyl alcohol	36 g acetic acid	210
Pineapple	ethyl formate	16 g ethyl alcohol	62 g butyric acid	121
Rum	ethyl butyrate	16 g ethyl alcohol	28 g formic acid	54
Apricot	amyl butyrate	22 g amyl alcohol	62 g butyric acid	158
Jasmine	methyl anthranilate	20 g methyl alcohol	82 g anthranilic acid	136

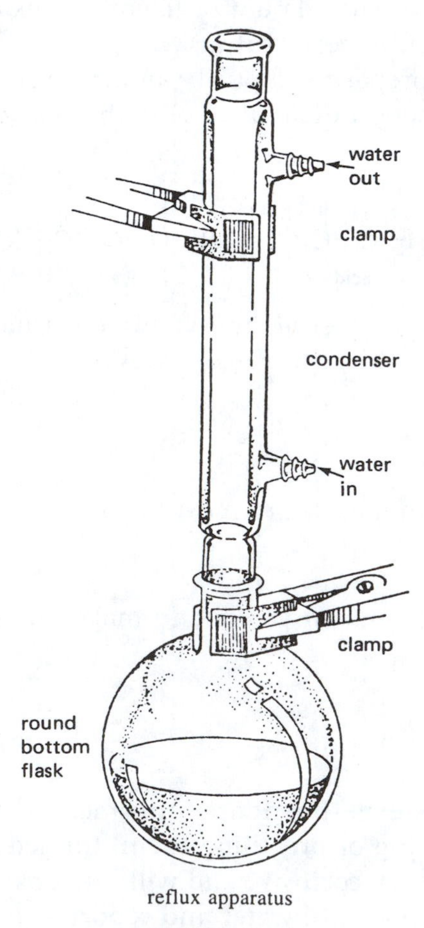

FIGURE 26-1

The reflux apparatus allows the mixture to boil without vapors escaping from the flask. Ask your instructor to check the apparatus before beginning. Adjust your flame so that a slow boil is maintained. Reflux for 1 hour.

Cool the ester mixture to room temperature and pour it into a separatory funnel. Add 50 mL of water, insert the stopper, invert the funnel (holding the stopper in place) and release the pressure by turning the stopcock. Repeat several times. This wash process removes the sulfuric acid catalyst and excess acids.

Separate the layers and discard the water layer. Return the organic ester mixture to the separatory funnel. Repeat the above wash procedure using sodium carbonate solution in place of water. Test with blue litmus paper to insure the organic layer is no longer acidic. The organic ester layer must be washed with fresh sodium carbonate solution until it is no longer acidic.

Pour the neutral organic layer into an Erlenmeyer flask and add about 2 g of anhydrous magnesium sulfate. Swirl gently for 2 minutes. Allow the ester to stand until liquid is clear. Add additional magnesium sulfate if necessary.

Assemble a distillation apparatus as shown in Figure 21-1, page 122. Make sure all joints are tight and that the condenser has water running through it. Carefully decant the ester into the distillation flask but leave the magnesium sulfate behind. Add 2-3 boiling chips. Ask your instructor to check the apparatus. Start the distillation and collect the fraction having the boiling point in the correct range (boiling point ±10°).

QUESTIONS

1. Which layer in the separatory funnel was the ester?

2. Why was the ester washed with sodium carbonate?

3. What happened when the ester mixture was washed with sodium carbonate?

4. Why was magnesium sulfate used?

5. Why do we isolate the ester over a boiling point range rather than at the correct boiling point?

Name ____________________ Class ______________ Room ___________

EXERCISE 27
SULFA DRUGS

INTRODUCTION

For centuries man has sought cures for the many diseases that affect him, but modern science has only recently been able to offer help. In the past 40 years doctors have used chemicals to destroy infectious microorganisms and cancerous cells with minimal effect upon the patient. This use of drugs to treat diseases is called chemotherapy and includes antibiotics and antimetabolites.

One group of antimetabolites are the sulfa drugs, discovered by Dr. Gerhard Domagk in 1935. Domagk used Prontosil, a red dye, to cure his daughter, who was dying of a streptococcal infection. The activity of Prontosil was due to its breakdown in the body to sulfanilamide. Sulfanilamide and related drugs are believed to be active against bacteria because they interfere with the synthesis of folic acid by the bacteria. Since we do not synthesize folic acid in our bodies, the sulfanilamide does not have this effect upon us. Sulfanilamide has been succeeded by a group of related compounds with equal or better therapeutic value and with less toxic side effects.

In this experiment you will prepare the original and simplest sulfa drug, sulfanilamide, from acetanilide. The reactions are

H O
N–C–CH_3

$\xrightarrow{HOSO_2Cl}$

H O
N–C–CH_3
SO_2Cl

$\xrightarrow{NH_4CH}$

H O
N–C–CH_3
SO_2NH_2

$\xrightarrow[(2)\ NaHCO_3]{(1)\ HCl}$

NH_2
SO_2NH_2

acetanilide

sulfanilamide

PRELAB STUDY QUIZ

1. What is chemotherapy?

2. Why must you be careful when using chlorosulfonic acid?

3. What process does sulfa drugs interfere with in bacteria?

SPECIAL HAZARD

Chlorosulfonic acid is very corrosive to the eyes and skin. Avoid contact and breathing. Chlorosulfonic acid may explode when added to water so avoid any water when using chlorosulfonic acid. Ammonium hydroxide is caustic and irritating to the eyes and mucous membranes. Avoid breathing and wash any affected areas with water. Hydrochloric acid will cause severe eye and skin burns. Wash any affected areas with large amounts of water and report any spills to the instructor.

PROCEDURE

CAUTION. Chlorosulfonic acid is an extremely hazardous compound. It reacts violently with water. All glassware must be completely dry. Chlorosulfonic acid also causes severe burns if spilled on the skin. If any should be spilled on your skin, wash with water and report immediately to your instructor. Do not pour any of this material into the sink. *Take only what you need.*

Add 13 mL chlorosulfonic acid to a dry 125-mL Erlenmeyer flask. Clamp the flask in an ice bath. Add 5 g of dry powdered acetanilide in 0.5-g portions. Mix with a stirring rod and keep the temperature below 20°C.

After most of the acetanilide has dissolved, warm the flask on a hot water bath for 20 minutes. Carefully pour the mixture onto 100 g of crushed ice. Stir the precipitate for a few minutes and then filter the product with a Buchner funnel. Use the product immediately in the next step since it reacts with water.

Transfer the crude product into a 125-mL Erlenmeyer flask and add 40 mL of concentrated ammonium hydroxide. Stir. Warm for 5 minutes on a steam bath. As the reaction proceeds, a thin paste of sulfanilamide is formed. Cool in an ice bath and isolate by suction filtration. You may stop here, until the next laboratory period, if necessary.

Place the crude product back in the 125-mL Erlenmeyer flask and add 20 mL 6 M hydrochloric acid. Warm the mixture over a very low flame and gently boil for 20 minutes. Cool the solution and add enough water to replace that lost by evaporation. (If a solid appears, reheat until no solid forms upon cooling.) If the solution is not colorless, add a small amount of activated carbon and filter by suction. Treat the filtrate with solid sodium bicarbonate, with stirring, until the solution is neutral. Induce precipitation by cooling in an ice bath and collect the solid sulfanilamide by suction filtration. Allow the product to dry until the next laboratory period. Weigh and determine its melting point.

DATA RECORD

Weight of sulfanilamide obtained ______________

Melting point of sulfanilamide ______________ °C

Melting point of sulfanilamide from literature ______________ °C

QUESTIONS

1. Penicillin is an antibiotic. How does it function?

2. List four diseases or disorders for which no chemical cure currently exists.

3. What does the addition of hydrochloric acid in the procedure do? Why is sodium bicarbonate added afterward?

Name ______________________ Class ____________ Room ____________

EXERCISE 28
ANALYSIS OF CIGARETTE SMOKE

INTRODUCTION

The average person breathes about 25 pounds of air each day to provide oxygen to the blood. The air that we breathe is usually not pure, but is filled with pollutants. Polluted air is a major factor in emphysema, which kills about 50,000 people in the United States each year, and in chronic bronchitis, which affects about one out of five men between the ages of 40 and 60. Yet with all the dirty air that we are forced to breathe, millions add to the problem by creating additional air pollution, by cigarette smoking. The Surgeon General of the United States has published research proving that smoking is a direct cause of lung cancer, and a warning to that effect is required on all cigarette packages and advertisements. The death rate from lung cancer is about ten times higher for cigarette smokers than for nonsmokers. Cigarette smoke irritates the bronchial tubes and renders them much more susceptible to disease. Some of the gases in cigarette smoke are carcinogenic. There is also strong evidence linking smoking with heart disease.

Solids and small particles are a major component of cigarette smoke. These inhaled particles pass through the body's defense mechanisms and destroy lung tissue. They range in size from submicron to several microns (10^{-6} meters), and are in part responsible for emphysema and chronic bronchitis. Some of these particles are mineral dust (such as lead), fly ash, and organic tars. In the lung they can irritate and destroy the alveoli (tiny sacs in the lung in which gas exchange occurs) and thus cause emphysema.

In the first part of the experiment you will determine the amount of solids in cigarette smoke from (a) a non-filter cigarette, (b) a filter-tip cigarette, (c) a filter-tip cigarette with the filter removed, (d) a "little-cigar" and (e) a low tar filter cigarette. In the second part of the experiment you will test the effectiveness of the filter in a filter-tip cigarette. This will be done by first determining the amount of solids that get through the filter when one-third of the cigarette is burned. This will be compared to the amount of solids that get through the filter when the cigarette is burned another one-third, and again to the solids passing through the filter when the last third of the cigarette is burned.

PRELAB STUDY QUIZ

1. About how much air is breathed each day by an average person?

2. What are the sizes of particles found in cigarette smoke?

3. Give 3 different examples of particles found in cigarette smoke.

4. How will you test the effectiveness of the filter in a cigarette?

PROCEDURE

1. Assemble the apparatus shown in Figure 28-1. Attach the rubber suction tubing to a water aspirator.
2. Weigh a piece of filter paper to the nearest 0.001 g. Place the filter paper in the apparatus as shown in Figure 28-1.
3. Weigh a non-filter-tip cigarette. Fit the cigarette into the glass tubing holder.
4. Turn on the water aspirator. Light the cigarette. To obtain good results, adjust the speed of the water aspirator so that the cigarette burns slowly over a period of 3 to 5 minutes. Rapid burning will produce poor results.
5. After the cigarette has burned as much as possible, turn off the aspirator and carefully extinguish the cigarette with a drop of water. Weigh the cigarette portion not burned. Remove and weigh the filter paper. Record the data.
6. Repeat the procedure for a filter-tip cigarette, a filter-tip cigarette with filter removed, a "little-cigar", and a low tar filtered cigarette. Record data.
7. Using a pencil or pen, mark off a filter-tip cigarette into 3 equal parts.
8. Repeat the same procedure but extinguish the cigarette after one third of the cigarette is consumed. Remove, weigh, and return the filter paper to the apparatus. Repeat for the next and final thirds.

Name ______________________ Class ______________ Room ______________

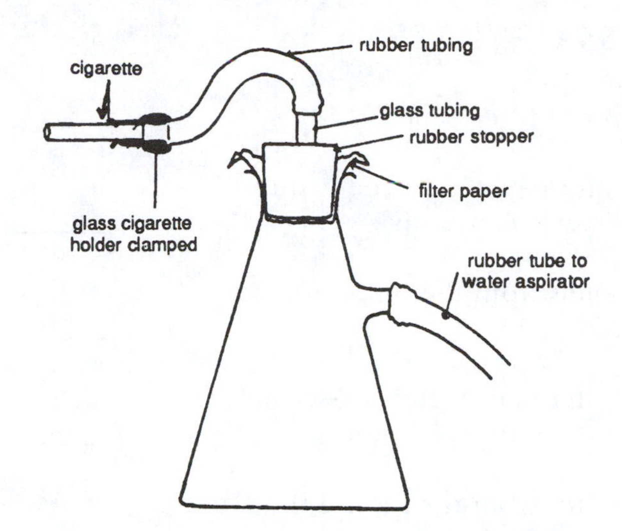

FIGURE 28-1 **Apparatus for cigarette smoke analysis**

DATA RECORD

SOLIDS IN VARIOUS CIGARETTES

	Non-filter Tip Cigarette	Filter-Tip Cigarette	Filter-Tip Cigarette Minus Filter	"Little Cigar"	Low-Tar Filter-Tip Cigarette
1. Initial weight of filter paper	______	______	______	______	______
2. Weight of filter paper after the cigarette is consumed	______	______	______	______	______
3. Weight of the solids on the filter paper (mg)	______	______	______	______	______
4. Initial weight of cigarette	______	______	______	______	______
5. Weight of the cigarette after burning	______	______	______	______	______
6. Loss of weight of cigarette after burning	______	______	______	______	______
7. Weight of solids collected per g of cigarette consumed (mg/g)	______	______	______	______	______

EFFECTIVENESS OF FILTERS

1. Initial weight of filter paper ______________________
2. Weight of filter paper after the first third of the cigarette is consumed. ______________________
3. Weight of solids from the first third of cigarette. ______________________
4. Weight of filter paper after the second third of the cigarette is consumed. ______________________
5. Weight of solids from the second third of cigarette. ______________________
6. Weight of filter paper after last third of cigarette is consumed. ______________________
7. Weight of solid from last third of cigarette. ______________________

QUESTIONS:

1. Which cigarettes produce the most solids? The least solids? Explain your results.

2. Which portion of the cigarette produced the most solids?

3. Which cigarette is the safest for smoking, based on your data?

Biochemistry Exercises

Name ________________ Class ________ Room ________

EXERCISE 29
CARBOHYDRATES AND LIPIDS

INTRODUCTION

Carbohydrates are an important part of our diet. They are a source of energy. They also are the principal structural material in plants. Carbohydrates contain alcohol groups and either aldehyde or ketone groups or yield compounds that do. The three major classes of carbohydrates are monosaccharides, disaccharides, and polysaccharides.

Monosaccharides include glucose, fructose, and galactose. They cannot be hydrolyzed to simpler sugars. They are good reducing agents. In addition, glucose and fructose can be fermented.

Disaccharides include sucrose, maltose, and lactose, each of which can be broken down into two monosaccharide molecules. Maltose and lactose are reducing sugars, and maltose and sucrose can be fermented.

Polysaccharides are polymers made up of many monosaccharide units. Glycogen, starch, and cellulose are polymers of glucose. None of the polysaccharides are water soluble. The hydrolysis of starch involves several steps: fist to dextrin, then to maltose, and finally to glucose. The progress of this reaction can be followed since iodine gives an intense blue-black color with starch and a red color with dextrin. Maltose and glucose are colorless in the presence of iodine.

Although there are many different kinds of lipids, they are all nonpolar, water-insoluble materials. Fats, one type of lipid, serve as stored energy, provide insulation for the body, and protect the body's internal organs. Cholesterol, another lipid, plays an important part in the production of hormones. (It is also thought to contribute to heart disease when it accumulates in the coronary arteries.)

Simple lipids are esters of fatty acids. Upon hydrolysis they yield a molecule of glycerol and three molecules of fatty acid. Both fats and oils are simple lipids, the difference being the number of double bonds (unsaturation) in the fatty acids. The more double bonds there are, the lower the melting point. Oils are therefore more unsaturated than fats. The removal of double bonds (by hydrogenation) will convert an oil into a fat.

Saponification is the hydrolysis of a fat or oil by a strong base. The products are glycerol and salts of the fatty acids rather than the fatty acids themselves. Sodium or potassium salts of fatty acids are used as soaps. Castile soap, for example, is made by the saponification of olive oil with sodium hydroxide.

PRELAB STUDY QUIZ

1. Which saccharides can be fermented?

2. Which saccharides can be reduced?

3. See if you can find out what differences between starch and cellulose allows you to digest one but not the other.

4. What are soaps?

5. What does "polyunsaturated" mean?

SPECIAL HAZARD

Methylene chloride is an anesthetic and narcotic if inhaled in large quantities. Avoid breathing and evaporate only in a hood. Hydrochloric acid will cause severe skin and eye damage. Wash any affected area with large amounts of water and report burns to the instructor.

PROCEDURE

Enter all results in the Data Record.

A. Reducing and Nonreducing Sugars

To detect the presence of a reducing sugar, the Benedict's test will be done on six carbohydrates: glucose, fructose, maltose, lactose, sucrose, and starch. Label six test tubes. Add 5 mL of Benedict's reagent to each and 10 drops of the appropriate sugar solution. Place the tubes in a beaker of boiling water for 5 to 7 minutes. Record your observations. Formation of a brick-red or yellow precipitate is the positive test for the presence of a reducing sugar.

B. Fermentation

Crush a piece of yeast cake (half a cake). Mix one half of the crushed yeast with 20 mL of glucose solution and the other half with 20 mL of lactose solution. Pour the solutions into two different fermentation tubes. Set the tubes aside in a warm location and observe occasionally. Record your observation. What is the gas being formed? What else is formed?

If fermentation tubes are not available, fill two test tubes with the sugar solutions. Put your thumb over the mouth of the tubes. Invert, and place them mouth downward into beakers containing a small amount of the same sugar solution and a small piece of yeast. Make sure the yeast is inside the test tube. The gas will collect at the top of the inverted test tube.

C. Hydrolysis of Carbohydrates

Add 5 mL of sucrose solution to one test tube and 5 mL of starch to another test tube. Add 2 drops of concentrated hydrochloric acid to each and boil carefully to 2 to 3 minutes. Cool to room temperature and add enough 10% sodium hydroxide to just turn the solution basic (until red litmus paper turns blue). Repeat the Benedict's test as described in Part A on each solution. Record your observations. Are the test results different from those you obtained in Part A?

D. Acrolein Test for Fats and Oils

Perform the acrolein test by preparing two test tubes as follows: Add 2 drops of glycerol to one tube and 2 drops of olive oil or substitute to the other tube. Add a few crystals of potassium bisulfate to each tube. Carefully heat both tubes until a color change begins. Cautiously smell the odors that arise from the two tubes. Record your observations.

E. Unsaturation Determination for Fats and Oils

Prepare four test tubes as follows: Place 5 mL of methylene chloride into each tube. Add 1 g of palmitic acid to the first tube, 1 g of oleic acid to the second, 1 mL of cottonseed oil to the third, and approximately 1 g of lard to the fourth. Dissolve the contents of each tube by tapping the tubes against the palm of your hand. Add 10 drops of Hanus iodine solution to each tube and record how long it takes for the color to disappear. Your instructor may substitute for some of the fats and oils listed. How is the rate of color disappearance related to the amount of unsaturation present?

F. Saponification: the Preparation of a Soap

Add 25 mL of alcohol to about 5 g of lard or vegetable oil in an evaporating dish. Add 1.5 mL of 50% sodium hydroxide solution (**Caution:** *Concentrated sodium hydroxide solution causes very severe skin and eye burns; if spillage occurs, clean up immediately. Wash immediately with plenty of water if the solution comes in contact with the skin.*) Stir the mixture and warm gently until a solid appears (about 5 minutes). (**Caution:** *Heating should be done with hotplates if available. If Bunsen burners are used, keep the flames low since alcohol will burn. Should the alcohol catch fire, cover the dish with your asbestos board until the fire is extinguished.*) After the alcohol has evaporated, allow the mixture to cool with stirring. Formation of a solid indicates the production of soap. Separate the solid from the rest of solution. (**Caution:** *Do not handle, as the soap still contains large amounts of sodium hydroxide.*) To remove the sodium hydroxide and glycerol from the soap wash it several times with 20-mL portions of a concentrated sodium chloride solution. Break up any large lumps of soaps to maximize the removal of glycerol and sodium hydroxide. After washing, dry your soap with paper towels or filter paper.

Dissolve a small piece of your soap in distilled water and check the pH with pH paper. Do the same thing with a commercial soap such as Ivory. Is soap acidic, basic, or neutral?

Take a small piece of your soap and wash your hands with it. Does it lather? Repeat using dilute salt water instead of tap water. Does it still lather? Try this test with the commercial soap.

DATA RECORD

A. Results of the Benedict's Test

Sugar	Color	Reducing or Nonreducing Sugar
Glucose		
Fructose		
Maltose		
Lactose		
Sucrose		
Starch		

B. Fermentation Results

Glucose

Lactose

Name ______________________ Class ____________ Room __________

What gas is formed?

Other products formed?

C. Hydrolysis Results

Starch

Sucrose

D. Acrolein Test Results

Glycerol

Olive oil or substitute

E. Unsaturation Test

Fat or Oil	Time for Color to Disappear
Palmitic acid	
Oleic acid	
Cottonseed oil	
Lard	

Explain the relationship between the amount of unsaturation present and the time it takes for the color to disappear.

Rank the fats and oils you tested in order from the most unsaturated to the least unsaturated.

F. Soap

pH of your soap ____________________

pH of commercial soap ____________________

Explain these pH results.

Did your soap lather?

Did it lather in salt water?

Compare your soap's lathering ability to that of the commercial soap.

Why are soap solutions basic?

Name ______________________ Class ____________ Room ____________

CASE HISTORY: DIABETES MELLITUS

Mrs. Pearson is a 47-year-old widow. She was admitted to the hospital after being examined at a local clinic. A blood sample showed 300 mg of glucose per 100 mL of whole blood. Mrs. Pearson could not give a reliable family history.

She is overweight (75 kg [165 lb]), and her physical examination revealed that over the past two months she has lost 7.0 kg (15.5 lb). The other symptoms she described are as follows: (1) a vigorous appetite, constantly hungry; (2) drinking unusual amounts of liquid daily; and (3) urinating large amounts frequently.

A glucose tolerance test was ordered. This showed a definite insulin response, and a diagnosis of diabetes mellitus was made. The patient responded to insulin therapy, dietary management, and antibiotics for urinary tract infection.

CLINICAL APPLICATIONS

1. Note the following laboratory reports for Mrs. Pearson below and on page 166.

MICHAEL REESE HOSPITAL AND MEDICAL CENTER

15875

UNIT RECORD NUMBER | DEPT. CODE 602 | ☒ PRIVATE OUTPATIENT | ☐ EMERGENCY ROOM | ☐ MANDEL CLINIC

DIAGNOSIS	COMPLETE URINALYSIS	PRE-OP	TOTAL CHARGE
Diabetes	☒ 111	☐	
SPECIMEN ☒ VOIDED ☐ CATH ☐ CLEAN CATCH ☐ MID-STREAM	DATE AND TIME COLLECTED		DATE EXAMINED
RESIDENT OR INTERN B. Cohen	REQUISITION PREPARED BY		TECH.

Amelia Pearson

CHART

LOW POWER FIELD			HIGH POWER FIELD	
CASTS		EPITHELIUM	RBC occ.	pH ☐ 113 5.0
HYALINE 7-15	FATTY	SQUAMOUS Mod.		SPECIFIC GRAVITY 114☐ 1.017
	WAXY		WBC 15-30	
GRANULAR FINE 0-2	DEGENERATED	ROUND Mod.	BACTERIA few	PROTEIN (Qual.) ☐ 115 0
			YEAST	SUGAR (Qual.) 116☐ ++++
COARSE rare	RBC	MUCUS THREADS few	TRICHOMONAS	ACETONE ☐ 117 SMALL
	WBC		AMORPHOUS	OCCULT BLOOD 118☐ 0
CYLINDROIDS	EPITH. CELL	DEBRIS	CRYSTALS	Bile 0

(a) Which tests did the doctor order?

(b) What are the normal ranges for each of these tests?

MICHAEL REESE HOSPITAL AND MEDICAL CENTER

214931 (R-1-76)

UNIT RECORD NUMBER	V	B	MO	DAY	YR	N	U	1	2	3	4	5	6	DEPT. CODE 603	☐ PRIVATE OUTPATIENT	☐ EMERGENCY ROOM	☐ MANDEL CLINIC

DIAGNOSIS: Diabetes

TOTAL CHARGE

SPECIMEN | DATE AND TIME COLLECTED | DATE EXAMINED

Amelia Pearson

RESIDENT OR INTERN: B. Cohen | REQUISITION PREPARED BY | TECH.

CHART

Test	Normal	Result	Test	Normal	Result	Test	Normal	Result	Test	Normal	Result
☒ SODIUM 111	138-142 Meq/l	142	☒ GLUCOSE 119 ☐ 2 Hr. Post Prandial	70-105 mg %	300	CALCIUM ☐ 124	9.5-10.5 mg %		AMYLASE ☐ 129	40-150 U.	
POTASSIUM 112 ☒	3.9-4.4 Meq/l	4.2	UREA N. 120 ☐	9-18 mg %		PHOSPHORUS 125 ☐	Ad.-2.5-4.0 Ch-4-7mg %		LIPASE 130 ☐	< 1.0 U.	
CHLORIDE ☒ 113	99-104 Meq/l	100	CREATININE ☐ 121	0.6-1.0 mg %		MAGNESIUM ☐ 126	1.8-2.3 mg %		Phenothiazines ☐ 191	Absent	
CO_2 CONTENT 114 ☒	26-29 mm/l	24	INSTRUCTIONS						Blood Ammonia 132 ☐	< 150 µg %	
BLOOD pH ☒ 115	7.38-7.42	7.35							SALICYLATE ☐ 133	Absen.	
BLOOD pCO_2 116 ☐	V=42-49mm A=38-41mm								BROMIDE 134 ☐	Absent	
BLOOD pO_2 ☐ 117	V=40 A=>93		☒ GLUCOSE TOL. 122	SEE SEPARATE REPORT		BILIRUBIN(T) ☐ 127	< 0.8 mg %		BARBITURATE ☐ 135	Absent	
ACETONE 118 ☐	< 9mg %		CREATINE 123 ☐	0.5-1.2 mg %		BILIRUBIN(G) 128 ☐			DORIDEN 136 ☐	Absent	

(c) Which results were outside the normal limits?

2. When insulin is used to treat diabetes mellitus, what determines the amount to be used? the frequency of administration?

3. How is insulin administered? Where?

4. What is a "diabetic coma"? What causes it and how should it be treated?

5. What problems are involved in healing cuts or wounds in diabetic patients?

6. What special problems do pregnant diabetics have?

7. Prepare a sample "diabetic diet." Would such a diet be good for all diabetic patients? Explain.

BASIC SCIENCE ASPECTS

1. What condition(s) might cause hyperglycemia? hypoglycemia?

2. Are diabetes mellitus and diabetes insipidus similar? Explain.

3. What could cause emotional glycosuria? What are its effects? How is it treated?

4. Why should insulin not be administered orally?

5. Describe a chemical test for the presence of sugar. Does a positive test not always indicate diabetes mellitus? Explain.

6. What is meant by the term "renal threshold"?

7. How is milligram percent (mg%) glucose related to percent by weight?

8. What are ketone bodies? How do they affect the pH of the blood? Do they always indicate diabetes mellitus? Explain.

9. What is ketonemia? ketonuria? ketosis?

10. Acetone on the breath of a person might indicate what condition? What is the source of the acetone?

11. What hormone(s) elevate blood sugar levels?

12. What hormone(s) lower blood sugar levels?

13. What is one unit of insulin equivalent to?

Name ______________________ Class ____________ Room ________

EXERCISE 30
PROTEINS AND NUCLEOPROTEINS

INTRODUCTION

Proteins are large molecules (molecular masses up to several million) found in all cells in the body. They are composed of approximately 20 simple building blocks called amino acids. The general formula for an amino acid is as follows:

$$\begin{array}{c} COOH \\ | \\ H_2N-C-H \\ | \\ R \end{array}$$

Proteins can be considered to be like very long words made up from a 20-letter, amino acid, alphabet. An example of a simple eight amino-acid word is shown below.

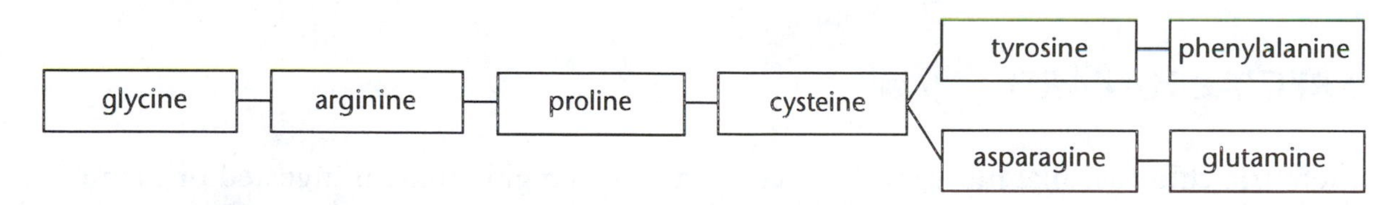

This is the symbolic structure for the hormone vasopressin.

Amino acids and proteins have many physical properties in common. Amino acids contain at least one acidic group and one basic amino group. Some, such as aspartic acid or lysine, contain more. Proteins also contain acidic and basic groups. When a large molecule, such as a protein, has a net ionic charge, its water solubility is enhanced. However, when the positive and negative charges are equal (the isoelectric point), the protein becomes much less soluble. You will study the acid-base properties of amino acids and proteins and their solubilities in this exercise.

In addition to the amino acids present in a protein, a protein is dependent upon its three-dimensional structure for its biologic activity. Destruction of a protein's amino acid sequence is called protein hydrolysis. Destruction of the three-dimensional structure is called denaturation. The three-dimensional structure is dependent upon hydrogen bonds, interionic bonds called salt bridges, and disulfide bonds. Agents that can interfere with these will denature a protein. You will examine several denaturing agents. See if you can figure out how they work.

Nucleoproteins are called conjugated proteins; that is, they contain both a protein and nonprotein part. Digestion separates these two parts. The nonprotein parts of RNA, called a nucleic acid, include a phosphate, a sugar (ribose), and an organic nitrogen base.

In this exercise, you will isolate the various components of a nucleoprotein derived from yeast RNA.

PRELAB STUDY QUIZ

1. What two functional groups are found in amino acids?

2. What is the function of vasopressin in the body?

3. What are the components of a nucleotide?

4. Under what conditions are proteins least soluble?

5. What does denaturation mean?

6. Give three ways to denature a protein.

SPECIAL HAZARD

Mercuric chloride, like most mercury compounds, is highly toxic if ingested or inhaled. Avoid skin contact or breathing the powder and wash hands after use. Do not pour mercury solutions or compounds down the sink or into waste cans. The instructor will provide a container for mercury disposal. Sodium hydroxide is very caustic and will cause skin and eye damage. Rinse your skin with water if it has a soapy feeling. Clean up all sodium hydroxide immediately. Report any skin burns to the instructor. Picric acid will explode by heat or percussion and will cause allergic reactions when handled. Avoid touching picric acid and keep it well away from any heat source.

PROCEDURE

Enter all observations and results on the Data Record.

Name ______________________ Class ____________ Room ____________

A. Acid-base Properties of Amino Acids and Proteins

Your instructor will have several different amino acids and proteins available. Determine their pH by using pH paper. Can you make a generalized statement concerning the acid-base properties of proteins?

B. Solubility

Put 2 mL of glycine solution into six test tubes. To the first, add 2 mL of cold water (pH 7); to the second, 2 mL of hot water (pH 7); to the third, 2 mL of physiologic saline solution (pH 7); to the fourth, 2 mL of 10% sodium hydroxide (pH > 14); to the fifth, 2 mL of 0.2% HCl (pH 1 to 2); and to the sixth, 2 mL of 0.5% sodium carbonate (pH 9 to 10). Record whether the glycine stays dissolved or not. Repeat with an egg albumin solution. Does the water temperature affect solubility? the presence of a salt? the pH?

C. Denaturing Protein

Put 5 mL of egg albumin into six text tubes. Heat the upper portion of the first tube until it boils. Add 5 mL of alcohol to the second. Add 2 mL of tannic or picric acid to the third. Add 3 mL of mercuric chloride to the fourth. Shake the fifth tube vigorously for 2 minutes. Record your observations.

D. Protein Identification-the Biuret Test

Place 5 mL of egg albumin mixture in a test tube. Then add 5 mL of 15% sodium hydroxide solution to the tube and mix by tapping the tube against the palm of your hand. Add 5 drops of dilute copper(II) sulfate solution to the tube. Observe and record the color. Repeat with glycine. A purple color indicates the presence of a protein.

E. Test of Protein, Ribose, Phosphate, and Purine in Yeast RNA

1. *Protein.* Test for the presence of protein by doing a biuret test on a small amount of yeast RNA suspended in 2 mL of water. Stopper and shake vigorously for a couple of minutes. Did the purple color indicating protein appear?
2. *Preparation of RNA Solution.* Put about 0.1 g of yeast RNA into a test tube, add 3 mL of water and 3 mL of 3 M sulfuric acid, and heat for 20 minutes in a hot water bath (80°C). Divide the resulting solution equally between three test tubes.
3. *Ribose.* To the first test tube of RNA solution, add 2 mL of water and 3 mL of Bial's orcinol reagent. Mix, and heat in a boiling water bath for 10 minutes. A green color indicates the presence of the sugar ribose.
4. *Purine.* To the second tube of RNA solution, add concentrated ammonium hydroxide until the solution is basic. (Check with litmus or pH paper.) Add 5 drops of silver nitrate solution. A precipitate indicates the presence of purine.

5. *Phosphate.* To the third tube of RNA solution, again make the solution basic with concentrated ammonium hydroxide and then turn it slightly acidic with nitric acid (check with litmus or pH paper). Add 2 mL of ammonium molybdate and warm. Formation of a yellow precipitate indicates the presence of phosphate.

DATA RECORD

A. pH of Amino Acid and Proteins

Amino Acid or Protein	pH
1.	
2.	
3.	
4.	
5.	
6.	
7.	

Statement concerning acid-base properties of proteins:

B. Solubility

Use S for soluble, I for insoluble in filling out the following table:

	Glycine	Egg Albumin
Cold water		
Hot water		
Saline solution		
Sodium hydroxide (pH > 14)		
Hydrochloric acid (pH 1–2)		
Sodium carbonate (pH 9–10)		

Does water temperature affect solubility?

Does the presence of a salt affect solubility?

Does the pH of the solution affect solubility?

C. Denaturing Observations

Heat

Alcohol

Tannic or picric acid

Mercuric chloride

Shaking

D. Biuret Test

Egg albumin

Glycine

E. Analysis of Yeast RNA

Protein

Ribose

Purine

Phosphate

QUESTIONS

1. Does water temperature affect solubility of protein? the presence of salt? the pH?

2. Why was picric acid once used in treating burns? What has replaced it?

3. Why are alcohol and mercuric chloride useful as disinfectants?

4. Why are egg whites used to treat heavy-metal poisonings? Why must an emetic be given soon afterward?

Name ______________________ Class __________ Room __________

EXERCISE 31
THE CHEMISTRY OF MILK

INTRODUCTION

Milk is the glandular secretion of the mammary glands of mammals for the nourishment of their young. During pregnancy increased amounts of estrogens cause the development of the mammary glands, so that they are ready for milk production at the end of pregnancy. Milk production is then initiated by the hormone prolactin, and release of milk from the breast by sucking is stimulated by the hormone oxytocin.

Since milk has been designed by nature for feeding newborns, it is an excellent nutrient. Milk contains fats, sugars, proteins, vitamin A, calcium, and phosphorus. However, milk is deficient in most of the B vitamins, vitamin C, vitamin D, copper, and iron. The composition of various sources of milk is shown in Table 31-1.

The three proteins in milk—casein, lactalbumin, and lactoglobulin—are sufficient to meet all the protein requirements of animals. Casein is the principal protein (about 80%), and it acts as a stabilizing agent for the emulsion of butterfat in water. Casein also contains most of the calcium and phosphate found in milk. The salt, calcium caseinate, is partially responsible for milk's white color.

TABLE 31-1

The Average Composition of Milk*

	Human Milk	Cow's Milk	Goat's Milk
Protein (%)	2	3	4
Carbohydrate (%)	7	5	5
Fat (%)	3	4	5
Calcium (%)	0.034	0.12	0.13
Phosphorus (%)	0.015	0.09	0.1
kcal/100 ml	63	69	98

*The remainder is primarily water.

The carbohydrate in milk is lactose. Lactose is found exclusively in milk. Lactose is made up of galactose and glucose and, since galactose is not found in significant amounts in the other body tissues, the mammary glands must isomerize glucose into galactose.

Upon standing, certain bacteria convert lactose into lactic acid. The milk is said to become sour. The formation of lactic acid causes the casein to precipitate from milk in the form of curds. This precipitated casein also contains fats and other substances from milk and is known as cottage cheese.

The vitamin content of milk supplies all the vitamins necessary for the infant of the species. Cow's milk, however, does not contain enough vitamin D for the winter months, when sunshine exposure is lessened, and has about one third the necessary amount of vitamin C. Pasteurization of cow's milk further reduces the vitamin C content.

In this exercise you will separate some of the constituents found in milk.

PRELAB STUDY QUIZ

1. What sugar is found in milk?

2. What two vitamins are not present in sufficient amounts in milk?

3. What constituent of milk contains most of the phosphorus and calcium?

4. What causes milk to sour?

5. What type of milk listed in Table 31-1 contains the most fat? sugar? highest energy content?

SPECIAL HAZARD

Diethyl ether is extremely fmallable and harmful to breath or touch. Keep away from all flames.

PROCEDURE

Enter all results on the Data Record.

A. Properties

Test milk with a piece of pH paper. Obtain a hydrometer and determine the density of milk.

B. Isolation of Casein

Pour 50 mL of whole milk into a 250-mL beaker. While stirring slowly, add 20 mL of glacial acetic acid. A precipitate of casein and milk fat should form. Filter the precipitate and save the filtrate (this liquid is called whey) for Part D. Wash the precipitate several times with water. Transfer the solid to a small beaker, cover with ethanol, and stir for a couple of minutes (this removes traces of water). Pour off the alcohol and cover the solid casein with about 10 mL of ether in order to extract the butterfat. (**Caution:** *No flames should be present since ether vapors are very flammable.*) Pour the ether extraction into an evaporating dish and add 10 mL more of ether to the casein. Combine the second ether extract with the first. The extractions separate the butterfat from the casein. Place the evaporating dish containing the ether aside, away from a flame, in the hood if possible, and allow the ether to evaporate. Save for Part C. The solid casein remaining in the beaker should be pressed dry between pieces of filter paper. Test the casein for protein by the biuret test (see page 171).

C. Test for Butterfat

After all the ether has evaporated, test the residue for fat by the acrolein test (see page 161).

D. Isolation of Lactalbumin and Lactoglobulin

Boil the filtrate obtained from the initial filtration of milk until lactalbumin and lactoglobulin coagulate. Filter and save the filtrate for Part E, F, and G. Test the solid for protein by the biuret test (see page 171).

E. Test for Lactose

Add 5 drops of the filtrate from Part D to 3 mL of Benedict's reagent and heat in a boiling water bath. A brown color indicates the presence of a sugar.

F. Test for Calcium Ion

Mix together 1 mL of ammonium oxalate solution and 1 mL of filtrate from Part D. A precipitate indicates the presence of calcium.

G. Test for Phosphate Ion

Mix together 4 drops of nitric acid, 2 mL of ammonium molybdate solution, and 3 mL of the filtrate from Part D and heat in a water bath. Formation of a precipitate indicates the presence of phosphate.

DATA RECORD

A. pH of milk ______________

Density of milk ______________

B. Does casein give a positive biuret test? ______________

C. Does butterfat give a positive acrolein test? ______________

D. Do lactoglobulin and lactalbumin give a positive biuret test? ______________

E. Do you get a positive test for sugar (lactose)? ______________

F. Do you get a positive test for calcium ion? ______________

G. Do you get a positive test for phosphate ion? ______________

QUESTIONS

1. Compare human milk with cow's milk in terms of nutritional value.

2. How does diet affect the contents of milk?

3. What effect might insecticides have on milk?

4. What effect might radioactive fallout have on milk?

Name ______________________ Class ____________ Room ____________

EXERCISE 32
ASCORBIC ACID CONTENT IN FRUIT JUICE

INTRODUCTION

Vitamin C (ascorbic acid) is a six-carbon compound structurally related to glucose and other hexoses. It is a water-soluble vitamin found in citrus fruits, the cabbage family,

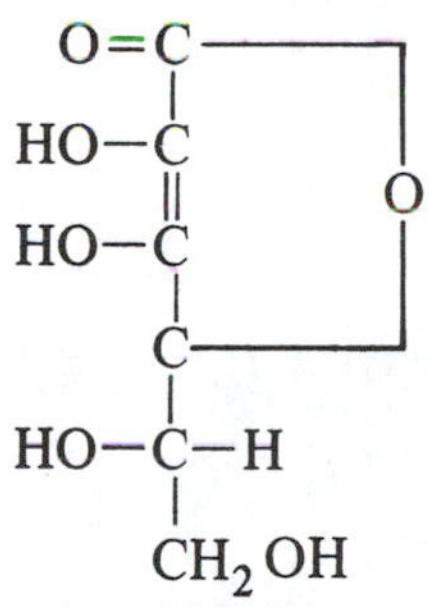

```
 O=C ─────┐
   |      |
HO─C      |
   ‖      O
HO─C      |
   |      |
   C──────┘
   |
HO─C─H
   |
   CH2OH
```

Ascorbic Acid

```
     OH
     |
   H─C ─────┐
     |      |
   H─C─OH   |
     |      |
  HO─C─H    O
     |      |
   H─C─OH   |
     |      |
   H─C──────┘
     |
     CH2OH
```

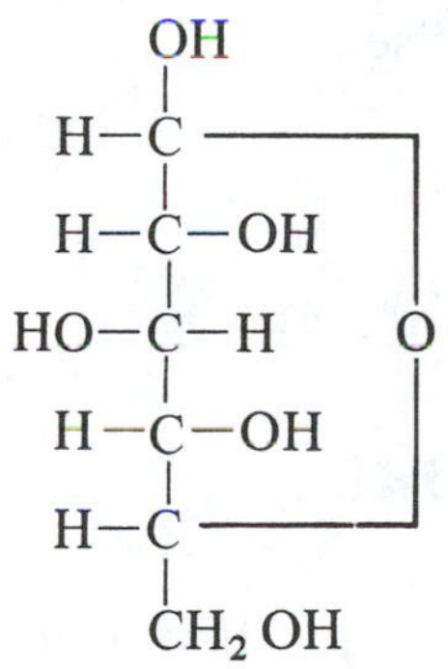

Glucose

tomatoes, liver, etc. It is essential in preventing scurvy, a disease that causes sore gums, loose teeth and hemorrhaging. In general, ascorbic acid functions in a variety of biochemical reactions that mostly involve oxidation. Although rather stable in acid solution, it is rapidly oxidized by air in neutral or alkaline solutions.

Man and other primates as well as the guinea pig are the only mammals known to be unable to make ascorbic acid. Without a dietary source of vitamin C, scurvy will develop. This most frequently happens among the elderly, alcoholics, drug addicts and undernourished infants. The daily intake must equal the amount lost by oxidation and excretion. About 3% of the body's store of ascorbic acid is lost per day. This produces a requirement of about 60 mg of vitamin C per day.

The ascorbic acid content of foods can be determined by oxidizing ascorbic acid to dehydroascorbic acid. In this experiment you will titrate a known volume of fruit juice with an iodine solution using starch as an indicator.

$$\text{ascorbic acid} + I_2 \xrightarrow{\text{starch}} \text{dehydroascorbic acid} + 2\ I^-$$

PRELAB STUDY QUIZ

1. In what kind of biochemical reactions is ascorbic acid involved?

2. What is the purpose of starch in the experiment?

3. What disease will Vitamin C help cure? What populations are likely to develop this disease?

PROCEDURE

You will be given a fruit juice to analyze for ascorbic acid content. Clean a 50 mL buret and fill it with a 0.0030 M iodine solution. Put 25 mL of 5% acetic acid solution into an Erlenmeyer flask. Pipet 20 mL of the fruit juice into the flask and add 3 mL of 1% starch solution. Titrate the fruit juice with the iodine solution until the blue-black color of the starch-iodine complex persists for 30 seconds. This is the end point and indicates that all the ascorbic acid has been converted to dehydroascorbic acid. Record the number of milliliters of iodine solution used. Repeat the titration with another sample of fruit juice. Your answers should agree within 5%. If they do not, run a third titration.

To calculate the milligrams of ascorbic acid in your sample use the following equation (concentrations of iodine solution other than 0.003 M iodine will have different values; consult your instructor).

$$\text{mg of ascorbic acid in sample} = \text{mL of } I_2 \text{ used} \times \frac{0.53 \text{ mg ascorbic acid}}{\text{mL of } I_2}$$

Name ______________________ Class ____________ Room ____________

DATA RECORD

	Titration 1	*Titration 2*	*Titration 3*
Initial buret level	______	______	______
Final buret level	______	______	______
mL of I_2 used	______	______	______
mg of ascorbic acid in sample	______	______	______
Average mg of ascorbic acid in sample		______	

QUESTIONS

1. Why would this experiment be difficult to do with grape juice?

2. Would you expect juices that have been exposed to the air for several days to have higher, lower, or about the same levels of ascorbic acid as unexposed fruit juices? Explain.

Name ______________________ Class ____________ Room ____________

EXERCISE 33
DIGESTION

INTRODUCTION

The process of digestion is essential for the absorption and utilization of food in the body. Digestion involves the hydrolysis of large molecules—carbohydrates, fats, and proteins—into small molecules—monosaccharides, fatty acids and glycerol, and amino acids. Some foods, such as inorganic salts, vitamins, monosaccharides, and water, do not require digestion. Digestion takes place in the mouth, the stomach, and the small intestine, with different enzymes functioning in each region to bring about the required hydrolytic reactions.

Digestion starts in the mouth, with chewing increasing the surface area and the enzyme salivary amylase (ptyalin) breaking down starches to dextrins and maltose. The enzyme maltase, secreted into the small intestine, completes the breakdown to glucose. The small intestine contains other enzymes that hydrolyze different saccharides.

The digestion of proteins to amino acids starts in the stomach. There the enzyme pepsin converts proteins to polypeptides. The conversion is competed in the small intestine, where a series of enzymes break down proteins and polypeptides to amino acids.

Almost all the fat digestion takes place in the small intestine. These fats are first emulsified by bile and then hydrolyzed to fatty acids and glycerol by the enzyme lipase (steapsin).

PRELAB STUDY QUIZ

1. What foods are digested in the mouth? the stomach? the small intestine?

2. Complete the following equations

(a)

$$\begin{array}{l} \qquad\qquad\quad H_2C-O-\overset{O}{\overset{\|}{C}}-C_{17}H_{35} \\ \qquad\qquad\qquad\ \ | \\ H_{39}C_{19}-\overset{O}{\overset{\|}{C}}-O-C-H \\ \qquad\qquad\qquad\ \ | \\ \qquad\qquad\quad H_2C-O-\overset{O}{\overset{\|}{C}}-C_{17}H_{29} \end{array} \xrightarrow[H_2O]{\text{steapsin}}$$

(b)

$$H_2N-\underset{CH_3}{\underset{|}{\overset{H}{\overset{|}{C}}}}-\overset{O}{\overset{\|}{C}}-\overset{H}{\overset{|}{N}}-\underset{H}{\underset{|}{\overset{H}{\overset{|}{C}}}}-\overset{O}{\overset{\|}{C}}-\overset{H}{\overset{|}{N}}-\underset{CH_2OH}{\underset{|}{\overset{H}{\overset{|}{C}}}}-\overset{O}{\overset{\|}{C}}-OH \xrightarrow[H_2O]{\text{tripeptidase}}$$

3. What useful digestive function does chewing do?

4. What is the digestive function of bile?

Name ______________________ Class ______________ Room ______________

SPECIAL HAZARD

Mercuric chloride is highly toxic if ingested or inhaled. Avoid skin contact or breathing the powder and wash hands immediately after use. Dispose of all solutions or solids containing mercury into a separately marked container.

PROCEDURE

Enter all results and observations in the Data Record.

A. Starch Digestion

Chew on something to stimulate the flow of saliva (rubber band, paraffin). Collect about 2 mL of your saliva (an amylase solution may be substituted for saliva) and add it to 15 mL of 1% cooked starch. Mix them and immediately remove a drop of this mixture. Record the time.

Put the drop into a test tube containing a drop of iodine solution. Warm the tube in a water bath (about 50°C) and record the color. Repeat this process at 1 minute intervals for 20 minutes. Record the time and your observations. The amount of undigested starch left is indicated by the blue color.

As soon as the contents of the tube no longer give a blue color with iodine solution, add 5 drops of the mixture to a test tube that contains 5 mL of Benedict's reagent. Place this tube into a beaker of boiling water for 5 minutes and observe. A green color is obtained for 0.25% glucose, a yellow-orange with 1% glucose, and a brick red color with greater than 2% glucose. Record your observations.

B. Protein Digestion

Place a very small piece of hard-boiled egg in each of five labeled test tubes. Add 5 mL of pepsin solution to the first tube, 5 mL of acidified pepsin to the second, 5 mL of basic trypsin to the third, 5 mL of distilled water to the fourth, and 5 mL of acidified pepsin plus mercuric chloride to the fifth. Place the tubes into a warm water bath (about 40°C) and observe for 1 hour. Record your observations.

Does protein hydrolysis depend on the pH? What effect does a mercury salt have on the hydrolysis? Record your answers.

C. Fat Digestion

Place 10 mL of milk into each of two numbered test tubes. Then add 2 mL of saturated litmus solution to each tube. If the tubes do not have a blue color, add dilute sodium carbonate, drop by drop, until the color turns blue. Add 2 mL of lipase (steapsin) solution to one of the tubes and 3 mL of distilled water to the other tube. Place the tubes in a beaker that contains water at about 40°C. Maintain this temperature. Observe the tubes over a period of 1 hour. Record any changes that take place.

DATA RECORD

A. Starch Digestion Observations

Time (minutes)	Color*
1	
2	
3	
4	
5	
6	
7	
8	
9	
10	
11	
12	
13	
14	
15	
16	
17	
18	
19	
20	

* Color obtained from iodine test.

Name ______________________ Class ____________ Room ____________

Results with Benedict's Solution

B. Protein Digestion Observations

Pepsin

Acidified pepsin

Basic trypsin

Distilled water

Acidified pepsin plus mercuric chloride

Does protein hydrolysis depend on pH?

What effect does a mercury salt have on protein hydrolysis?

C. Fat Digestion Observations

QUESTIONS

1. List the enzymes involved in the digestion of carbohydrate. Where is each found and what is the function?

2. How may an enzyme be deactivated?

3. What stimulates the flow of saliva? What inhibits it?

4. Describe the test for the presence of starch.

5. Does any digestion of fat take place in the stomach? Explain.

6. Where does the rebuilding of protein from amino acids take place in the body?

Name ______________________ Class ____________ Room ____________

EXERCISE 34
ENZYME CATALYSIS

INTRODUCTION

Reactions that occur in living systems involve common organic groups such as alcohol, ester, and amide. Many reactions involving of living systems (*in vivo*) can be done in laboratory glassware (*in vitro*) to make the same products. Reactions conditions must be very similar if the results from *in vivo* and *in vitro* experiments are to be compared.

Enzymes allow biological reactions that would normally require elevated temperatures or long reactions times to occur *in vivo*. Enzymes can also be used to to study reactions *in vitro*. Enzymes catalyze specific reactions of certain substrates. Some enzymes catalyze one type of reaction with a variety of substrates, while other enzymes catalyze only a single reaction for a single substrate.

Enzyme activity depends on the three-dimensional structure of the enzyme. Salt bridges, hydrogen bonds, disulfide bonds and hydrophobic interactions are some of the ways enzymes use to maintain their shape. Any chemical which can change these bonds will alter the shape and deactivate or denature an enzyme. Heat, metal ions, pH, and solvent can all bring about changes in enzyme activity.

Many enzymes are affected by compounds known as inhibitors. Competitive inhibitors are compounds that prevent enzyme-substrate interaction by occupying the active site of an enzyme. Noncompetitive inhibitors bind at sites other than the active site but produce enough change in the shape of the active site to change the enzyme activity.

You will study the effect of the enzyme catalase on the decomposition of hydrogen peroxide.

$$2H_2O_2 \longrightarrow 2\ H_2O + O_2$$

Hydrogen peroxide is produced in some cells and is toxic if it accumulates. The enzyme catalase is found in the red blood cell, liver and many plants. You will compare the results obtained by changing the pH, temperature, adding copper sulfate, an inhibitor, and denaturing the enzyme.

PRELAB STUDY QUIZ

1. What is the difference between an *in vitro* and *in vivo* reaction?

2. What enzymes are you studying in this experiment? What reaction does it catalyze?

3. How will you denature the enzyme?

PROCEDURE

Enter all results and observations on the Data Record.

Assemble the apparatus shown in Figure 34-1. Plastic tubes are preferable to glass tubes. Be very careful when inserting glass tubing into rubber stoppers (see page 8, Figure 10). The oxygen is bubbled into the graduated cylinder, displacing the water. The volume of oxygen generated is measured by the graduated cylinder. Clamp the apparatus securely so that the rubber stopper is immediately inserted into the test tube after mixing enzyme and substrate.

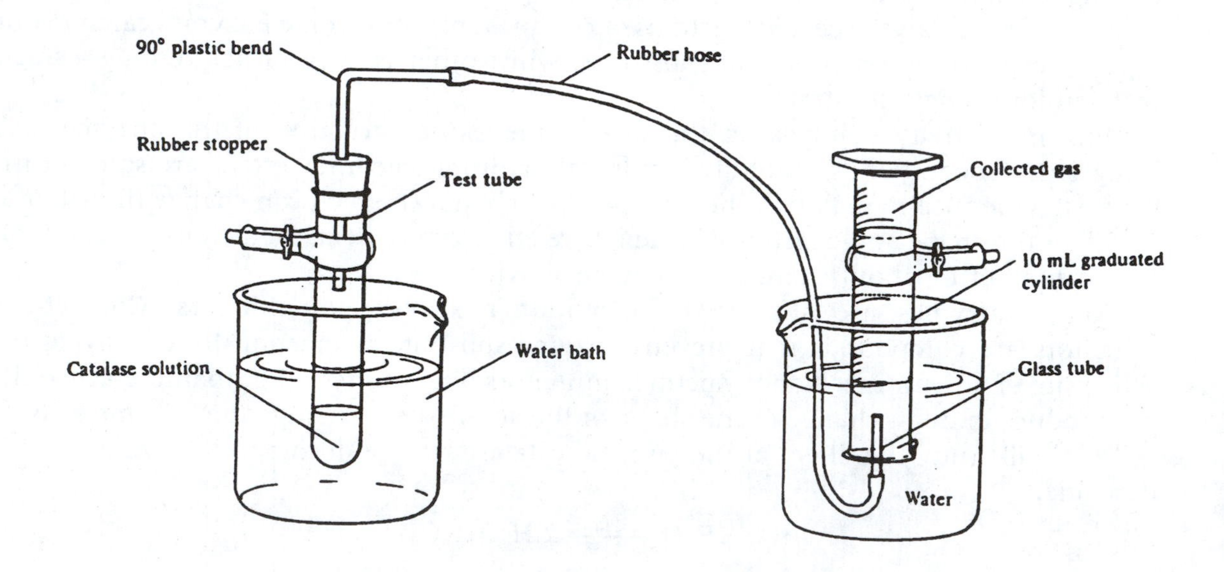

FIGURE 34-1

Name ______________________ Class ____________ Room ____________

Part A—Effect of Temperature

You will need six clean test tubes. Place two test tubes into an ice water bath. Add 2.0 mL of 3% hydrogen peroxide by pipet or buret to one test tube and two mL of the catalase solution to the other test tube. Allow the two solutions to cool for about five minutes. Now quickly add the catalase solution to the hydrogen peroxide solution, stopper quickly as shown in Figure 34-1, swirl the mixture, and note the time to the nearest second. Record the time needed to produce 5 mL of oxygen.

Repeat this procedure with water baths at 20 °C and 37 °C. Record times needed to produce 5 mL of oxygen and calculate the rate of oxygen production in mL of oxygen/sec.

Part B—Effect of pH

You will need four clean test tubes for this part. Add 2.0 mL of catalase solution and five drops of 0.1 M HCl to a test tube. In a different test tube, add 2.0 mL of 3% hydrogen peroxide. Place both test tubes in the 20 °C water bath and allow to stand for five minutes. Quickly add the hydrogen peroxide to catalase solution and collect 5 mL of oxygen as outlined in Part A. Repeat this procedure using five drops of 0.1 M NaOH. Record your results and calculate the oxygen production in mL of oxygen/sec. Compare the acid solution (pH=2), the alkaline solution (pH=12) and the solution from Part A (20 °C, pH=7).

Part C—Effect of an Inhibitor

You will need two clean test tubes for this part. Add 2.0 mL of catalase solution and four drops of 0.1 M $CuSO_4$ to a test tube. In a separate test tube add 2.0 mL of 0.1 M $CuSO_4$ and place both test tubes in the 20 °C bath for five minutes. Quickly add the hydrogen peroxide to the catalase solution and collect 5 mL of oxygen as before. Record your observations and calculate the rate of oxygen production.

Part D—Effects of Heat

You will need two clean test tubes for this part. Add 2.0 mL of catalase solution to the test tube and place the test tube in boiling water for five minutes. Cool the catalase solution to room temperature and place it along with a test tube containing 2.0 mL of hydrogen peroxide solution into the 20 °C water bath for five minutes. Quickly add the hydrogen peroxide solution to the catalase solution and collect oxygen as before. Record your results and calculate the rate of oxygen production.

Part E—Results

Explain the effect of different temperatures, pH, inhibitor, and heat on your results.

DATA RECORD

Part A—Effect of Temperature

Temperature (°C)	Time (sec)	Volume of oxygen (mL)	Rate mL/sec

Part B—Effect of pH at 20 °C

pH	Time (sec)	Volume of oxygen (mL)	Rate mL/sec
2			
7			
12			

Part C—Effect of inhibitor at 20 °C

Inhibitor	Time (sec)	Volume of oxygen (mL)	Rate mL/sec
$CuSO_4$			
None			

Part D—Denaturation

State of Enzyme	Time (sec)	Volume of oxygen (mL)	Rate mL/sec
normal			
denatured			

Name ______________________________ Class ______________ Room ______________

EXERCISE 35
CHEMISTRY OF A PHYSIOLOGICAL BUFFER SYSTEM
THE DIHYDROGEN PHOSPHATE/MONOHYDROGEN PHOSPHATE BUFFER

INTRODUCTION

The concept of buffers was introduced in Exercise 16. Under normal conditions the extracellular concentration of H^+ is 4×10^{-8} moles/liter. This is about one million times less than the concentration of other important ions such as Na^+, K^+, Cl^-, and HCO_3^-. Yet only very small variations in the H^+ concentration are consistent with life. Part of the reason is that H^+ is extremely reactive and has a profound influence on many physiological processes, such as respiration and glycosis. How does the body maintain a proper H^+ concentration? There are three different ways:

1. Chemical buffering
2. Respiratory control of CO_2 levels in the blood
3. Renal excretion of H^+

This experiment looks at one of the important buffer systems in the body: $H_2PO_4^-/HPO_4^{2-}$.

The following equation indicates the ionization of a weak acid anion, $H_2PO_4^-$.

$$H_2PO_4^- \rightleftharpoons HPO_4^{2-} + H^+$$

An equilibrium expression can be written for this reaction as follows:

$$K = \frac{[HPO_4^{2-}] + [H^+]}{[H_2PO_4^-]}$$

The formulas in brackets are the molar concentrations of the various ions and K is an ionization constant characteristic for the acid.

If we rearrange this equation and take the logarithm of both sides, we have the following expression known as the Henderson-Hasselbach equation. (Your instructor may want to go through the individual steps in its derivation.)

$$pH = pK + \log \frac{[HPO_4^{2-}]}{[H_2PO_4^-]}$$

When $[HPO_4^{2-}] = [H_2PO_4^-]$, then $\log [HPO_4^{2-}]/[H_2PO_4^-] = 0$ and pH = p*K*. Most buffers work best when p*K* = pH because the pH changes are much smaller when acids or bases are added. You will determine the p*K* for the ionization of $H_2PO_4^-$ by measuring the pH of various $[HPO_4^{2-}]/[H_2PO_4^-]$ mixtures and using the Henderson-Hasselbach equation. You will then determine accurately the pH variations in buffered and unbuffered solutions when acids and bases are added.

PRELAB STUDY QUIZ

1. When does pH = p*K* for a buffer solution?

2. Show how the $[H_2PO_4^-]/[HPO_4^{2-}]$ buffer system works when a base B or an acid H^+ is added.

3. What other buffer systems have physiological importance in the body?

4. Under what conditions are buffers most effective?

SPECIAL HAZARD

Hydrochloric acid and sodium hydroxide will cause severe eye and skin damage. Wash any affected area with large amounts of water and clean up any spills. Report any burns to the instructor.

PROCEDURE

Obtain 600 mL each of 0.1 M Na_2HPO_4 and 0.1 M NaH_2PO_4. Measure the pH of each Na_2HPO_4/NaH_2PO_4 mixture shown in Table 35-1, using a pH meter. (Your instructor will

Name ______________________ Class ______________ Room ____________

show you how to use a pH meter if you are unfamiliar with this piece of equipment.) Be sure to rinse the electrodes between pH determinations.

TABLE 35-1

Na_2HPO_4/NaH_2PO_4 Mixtures

	(1)	(2)	(3)	(4)	(5)	(6)	(7)	(8)	(9)	(10)	(11)
ml 0.1 M Na_2HPO_4	100	90	80	75	60	50	40	25	20	10	0
ml 0.1 M NaH_2PO_4	0	10	20	25	40	50	60	75	80	90	100

Calculate the concentrations of each mixture. *Remember the dilution involved.* Calculate the ratio of HPO_4^{2-} $H_2PO_4^-$ for each mixture. Convert these ratios to logarithms. (Your instructor may wish to provide these logarithms for you.) Calculate the p*K* for each mixture using the following equation:

$$pK = pH - \log \frac{[HPO_4^{2-}]}{[H_2PO_4^-]}$$

To illustrate the influence of a buffer on pH, you will add small amounts of acid and base to separate containers of the following and measure the pH; distilled water; 0.1 M NaH_2PO_4; 0.1 M Na_2HPO_4; and the $[HPO_4^{2-}]/[H_2PO_4^-]$ approximately equal to that in the bloodstream.

Place 50 mL of each of the preceding liquids in 100-mL beakers and measure the pH. Add 1 drop of concentrated HCl to each beaker and measure the pH. Add 3 drops of concentrated HCl and redetermine the pH in each beaker. Finally, add 5 drops of concentrated HCl and again measure the pH.

Repeat this part of the experiment using concentrated NaOH in place of HCl.

DATA RECORD

	(1)	(2)	(3)	(4)	(5)	(6)	(7)	(8)	(9)	(10)	(11)
ml 0.1 M Na_2HPO_4	100	90	80	75	60	50	40	25	20	10	0
ml 0.1 M NaH_2PO_4	0	10	20	25	40	50	60	75	80	90	100
pH of mixture											
Concentration of Na_2HPO_4											
Concentration of NaH_2PO_4											
$[HPO_4^{2-}]/[H_2PO_4^-]$											
p*K*											

	Distilled Water	0.1 M Na_2HPO_4	0.1 M NaH_2PO_4	Na_2HPO_4/NaH_2PO_4 Mixture
Original pH				
pH after addition of				
1 drop conc HCl				
3 drops conc HCl				
5 drops conc HCl				
1 drop conc NaOH				
3 drops conc NaOH				
5 drops conc NaOH				

QUESTIONS

1. What happens to the pH as the percentage of Na_2HPO_4 in the mixtures shown in the first table decreases? Why?

2. Where in the body does the $[H_2PO_4^-]/[HPO_4^{2-}]$ buffer system have important physiological roles?

3. Although the most effective buffering action occurs when the pH is near the p*K*, the p*K* for HCO_3^-/H_2CO_3 system is 1.3 units less than the pH. This means that the $[HCO_3^-]$ is about 20 times larger than $[H_2CO_3]$. Why is this system still an effective buffer?

4. For which $[HPO_4^{2-}]/[H_2PO_4^-]$ is p*K* directly obtainable from pH?

Name ______________________ Class ____________ Room ____________

5. What happens to the pH as the percentage of Na_2HPO_4 decreases? Why?

6. Which $[HPO_4^{2-}]/[H_2PO_4^-]$ is approximately equal to that in the blood?

NOTES

NOTES

NOTES

NOTES

NOTES